Carlos Alberto Amaya Corredor
Carolina Hernández Contreras
Alba J Vargas Buitrago

Environmental Management for the Construction of Sustainable Development

Carlos Alberto Amaya Corredor
Carolina Hernández Contreras
Alba J Vargas Buitrago

Environmental Management for the Construction of Sustainable Development

Compilation of research experiences of teachers and students, in the training of Environmental Engineering

ScienciaScripts

Imprint
Any brand names and product names mentioned in this book are subject to trademark, brand or patent protection and are trademarks or registered trademarks of their respective holders. The use of brand names, product names, common names, trade names, product descriptions etc. even without a particular marking in this work is in no way to be construed to mean that such names may be regarded as unrestricted in respect of trademark and brand protection legislation and could thus be used by anyone.

Cover image: www.ingimage.com

This book is a translation from the original published under ISBN 978-613-9-40253-3.

Publisher:
Sciencia Scripts
is a trademark of
Dodo Books Indian Ocean Ltd. and OmniScriptum S.R.L publishing group

120 High Road, East Finchley, London, N2 9ED, United Kingdom
Str. Armeneasca 28/1, office 1, Chisinau MD-2012, Republic of Moldova, Europe
Printed at: see last page
ISBN: 978-620-7-63768-3

GESTIÓN AMBIENTAL PARA LA CONSTRUCCIÓN DEL DESARROLLO SOSTENIBLE
Recopilación de experiencias de docentes y estudiantes
CO2
uts
Ambiental

AUTHORS

Natalia Alexandra Bohórquez Toledo

Chemistry, M.Sc. in chemistry

Research professor of the GRIIV green engineering research group, assigned to the Environmental Engineering program of the Technological Units of Santander

nbohorquez@correo.uts.edu.co

Alba Josefa Vargas Buitrago

Chemical Engineer, Environmental Engineering Specialist, M.Sc. in Environmental Sciences and Technologies

Research professor of the GIECSA Research Group, assigned to the Environmental Engineering program of the Technological Units of Santander

avargas@correo.uts.edu.co

Carolina Hernandez Contreras

Biologist. M.Sc. in Environmental Sciences and Technologies

Research professor of the GIECSA Research Group, assigned to the Environmental Engineering program of the Technological Units of Santander

chernandez@correo.uts.edu.co

Carlos Alberto Amaya Corredor

Cadastral and Geodesta Engineer, Specialist in Planning for Environmental Education, Master in Environmental Management and Audits, M.Sc. Sustainable Development and Environment

Research professor of the GIECSA Research Group, assigned to the Environmental Engineering program of the Technological Units of Santander

camaya@correo.uts.edu.co

INDEX

MISSION

The Green Engineering Research Group – GRIIV, is a research group whose main purpose is to generate knowledge in the different forms of resource treatment (Air, Soil and Water) to search for a new eco-efficient and sustainable methodology, in addition to carrying out studies ecotoxicological studies associated with the adverse effects caused by primary and secondary pollutants in natural resources and studies related to the application of chemical concepts aimed at generating eco-friendly impacts in industrial processes or recovery of different resources. In addition, its strength is projects focused on the external sector, such as the formulation and implementation of PGIR and the evaluation of environmental impacts. For this purpose, it focuses its activity on the training of Environmental Technologists and Engineers with a high degree of scientific training, proposing and developing a line of research in green processes, which responds to the solution of the country's needs in the industry of different processes, in the environmental and agroindustrial part.

VISION

In 2024, the GRIIV group will be a recognized research group and will maintain its classification in the national science and technology system in category C and will carry out strong investigative work to rise to category B, through the formulation and execution of projects. of research and development, of an innovative nature, aimed at analyzing and evaluating environmental problems and our surroundings that allow research and development projects to be carried out on the environmental part, working hand in hand with other research centers and groups and with the industry, both national and international. In order to advance in the development of methodologies that are environmentally friendly.

MISSION

Generate and apply knowledge, from the formulation and development of research projects, in the processes of planning, management and use of the natural resources of the Andean region and the environmental services they generate for sustainable human and social development.

VISION

By 2025, the GIECSA group will have recognition and classification within the national science and technology system; from the formulation, development and execution of research projects, aimed at the sustainable management and use of ecosystems and the environmental services that they provide; contributing to the environmental management of the territory in the area of influence of the Technological Units of Santander, achieving recognition as an agent of change that contributes to sustainable development, in the face of the new climate change and post-conflict scenarios that the country must face.

PRÓLOGO
uts Unidades Tecnológicas de Santander
¡Lo hacemos posible!
UDI UNIVERSIDAD DE INVESTIGACIÓN Y DESARROLLO

FOREWORD

Sustainability became a call for responsibility, from the international perspective, for the direction of individual actions in the search for common well-being from individual interests and achievements.

The path begun in 1987 by the Brundtland Commission, by giving it a surname when understanding Development, has little to do, for many it has been well traveled, for many others it has been poorly traveled, and for many others it has not yet been taken, but in any view, it has begun a path that advances towards common well-being, better forms of social relations, respect of each man for nature and understanding that individual actions have general consequences, so in every aspect in which I know it is part of Development, the surname Sostenible, or Sustainable (without discussing the semantic significance of each Hispanic word, but focused on achieving new and better results), must be the primary focus, directing actions and efforts for the common good, of our current generation and of the generations that will receive a neutral planetary legacy.

Sustainable development integrates in balance: environmental aspects, as a primary source of all resources usable by man, social aspects, such as the specialization of human actions and relationships to provide quality of life, economic aspects, such as the guidelines for the exchange of goods and services, markets and money flows as an exchange mechanism. The interaction of these elements must build viable, livable and equitable conditions for all, in the consolidation of quality of life, access and enjoyment of money flows, markets and goods and services.

These aspects and their relationships are present in all the actions that are generated daily, intentionally when the actions are planned or spontaneously when the initiatives flow towards good results, but they can also be absent or weakened when the actions are intended for biased benefits and in ignorance of the biosystemic reality in which we are all immersed. When this happens, a visibility strategy of the importance of sustainability helps to demonstrate and value the contributions to sustainability that all actions and processes can generate.

With the impetus of the Brundtland commission, in 1992, the world was summoned to the UN summit in Rio de Janeiro, in which multiple aspects that promoted well-being in the world were contemplated for the arrival of the new millennium, for which The sum of effort would be the fundamental basis of any initiative. RIO92 brought together world political leadership, proactive business initiatives for the planet and social movements for human and planetary well-being, generating the Agenda of the new millennium with what should be the transformative stakes of that moment. For the new millennium, the UN2105 summit in Paris marks the redesign of the sustainable development approach, by assuming the 2030 Agenda for development and within it, the Sustainable Development Goals (UN, 2015), as the common commitments to transform the world and achieve human and social well-being in balance with nature. In this approach, society, all economic actors and environmental processes must be integrated with more strength and intention to achieve sustainable, long-lasting and efficient results in the well-being of life in the world.

Strategic management and planning is a mechanism with great potential due to its integrative capacity of all parts, so its belonging to sustainability is undeniable by promoting the balance and well-being of all parts, focusing efforts on results, contemplating and maximizing the resources and promote human well-being as a key aspect of productivity (Gómez, 2018). For its part, environmental management contemplates the use of natural goods and services to meet human needs, regulating their waste or disposal so as not to deteriorate the living environment and, on the contrary, improve coexistence after enjoying the environment, protection and conservation of nature and the quality of life of all inhabitants.

Companies and organizations have key aspects to consolidate their sustainability if they integrate strategic management, for the productive efficiency of their processes and environmental management, such as guidelines for efficiency in their resources and regulation of their effects, thereby promoting their social and commercial positioning. by generating results that demonstrate concern for their environment, in terms of efficient products for social needs, that do not affect living environments and that are durable over time (Isaac, Gómez, & Díaz, 2017), aspects that result in benefits evident in economic solidity, respect, valuation and identity with the environment and respect and understanding of the dimension and social scope of their activities.

Describing them from their relationship to the Sustainable Development Goals makes it clear that the research processes of universities focus on the well-being stakes that interest the current world and the near horizon in the decade that advances until reaching the year 2030. Strategic management and environmental management, related to potential in companies, leaves ideas of how the productive sector has the possibilities of positively affecting its environment, after recognizing its resources and regulating its waste, approaching the population recognizing the common environment and offering products designed and pertinent to the realities of the population and ultimately, as aspects that strengthen and increase their economic balance, in which their profits and returns drive their commitment to sustainability as an integral contribution for social good.

Carlos Alberto Amaya Corredor

GIECSA Group, UTS

INTRODUCCIÓN
Unidades Tecnológicas de Santander
¡Lo hacemos posible!
UNIVERSIDAD DE INVESTIGACIÓN Y DESARROLLO

INTRODUCTION

Today we are all experiencing conditions that show us changes in the global climate, increased solar radiation, longer periods of drought, heavier rains, more intense storms, stronger environmental damage, loss of infrastructure, weakening of resource sources for productive systems, which have also been evidenced in alterations in the economic dynamics and life dynamics of society.

Since the Rio '92 Earth Summit, the creation of the Intergovernmental Panel on Climate Change (IPCC) allowed the topic of climate change to be put into global discussion, generating defenders and detractors, but allowing it to be recognized and learn to make global, regional or local decisions that allow generating actions to intervene in the damages that society recognizes coming from the climate.

In the fifth climate change assessment report (IPCC, 2014), the IPCC defined the existence of risks generated by the combination of factors: exposure, vulnerability and exposed danger of territories and communities. From these factors it is possible to identify the impacts generated by climate change, which affect the climate and socioeconomic processes. On socioeconomic processes, it affects productive dynamics, the dynamics of adaptation and mitigation measures and governance models, which are generators of atmospheric emissions and changes in land use, which degenerate into climate changes of anthropogenic origin and consequently alterations of natural climate variabilities.

The effects of climate change have been considered by the IPCC at a global level and countries have made efforts to measure it in their territories. Pabón (Pabón, 2018)shows how in Colombia it has been possible to identify the conditions of climate change, according to the different geographical regions and according to the variation of conditions such as temperature and precipitation. For his part (Costa, 2007), Costa makes the same description, also taking into account political, economic and social conditions, thereby giving a special condition to the way in which the country must face the challenges that climate change imposes on it.

In Colombia, the city of Bucaramanga, since 2010, began its participation in the Inter-American Development Bank (IDB) program for emerging and sustainable cities in Latin America, with which formulating and implementing urban sustainability strategies has been a condition of the governments in power, to comply with the requirements of this IDB program, and transform urban environments in favor of the quality of life of its inhabitants. Through IDEAM and the Ministry of Environment, in 2015, the departmental plan for adaptation and mitigation to climate change, Santander 2030 (MINAMBIENTE, 2017), was formulated, which recognizes the need to strengthen the processes of the city of Bucaramanga, to address this condition. .

Although the Santander 2030 document does not specify the processes or strategies to follow for the adaptation and mitigation of climate change in the capital of the department, the references to be met are established, from which it is possible to build applicable alternatives for the reality of the city.

From the UTS, in the construction of the Adaptation and Mitigation plan to climate change of the UTS, (AMAYA & HERNANDEZ, 2018)strategic action programs were generated to intervene in different spaces of presence of the institution, one program is UTS Solidaria, with a line of action on Climate Change: Promotion of UTS commitment in the area of influence of the institution, proposing as results: Institutional contribution to social awareness about climate change and Increase in institutional results projected in the communities of influence of the UTS.

As a result of this recognition of information, from the general to the particular, it has been possible to propose a set of strategies that promote the adaptation and mitigation of a sector of the city, Ciudadela Real de Minas, so that experiences are established that serve as a replica extended throughout the city, and advance in building city reconstruction mechanisms, initially from citizens transforming their lifestyles or customs of life, to planning, foresight and city infrastructure.

CAPITULO 1.
EDUCACIÓN AMBIENTAL, SOLUCIÓN SOSTENIBLE A LA PROBLEMÁTICA DE LA GESTIÓN DE LOS RESIDUOS SÓLIDOS

CHAPTER 1.
ENVIRONMENTAL EDUCATION, SUSTAINABLE SOLUTION TO THE PROBLEM OF SOLID WASTE MANAGEMENT

Alba Josefa Vargas Buitrago

Chemical Engineer, Environmental Engineering Specialist, M.Sc. in Environmental Sciences and Technologies

Research professor of the GIECSA Research Group, assigned to the Environmental Engineering program of the Technological Units of Santander

avargas@correo.uts.edu.co

ODS: 

SUMMARY

Market places generate considerable volumes of solid waste, which contribute significantly to environmental pollution, mainly due to inadequate management, which causes significant environmental impacts. The disposal of organic waste in the Bucaramanga city landfill is estimated at 88.18% of the total value, of which 33.33% is food waste and a good part of this waste is of plant origin according to the Bucaramanga mayor's office. The solid waste of plant origin that is produced in the squares comes from the waste of various types of fruits, legumes and vegetables, among others. The objective of this research was to apply an environmental sustainability strategy using environmental education as a tool to transform society and generate habits in the population of the area of influence of the square and in this way contribute to the sustainable development objectives SDG. 11 and SDG 12, to make cities and communities sustainable, as well as responsible production and consumption by the community. The case study was developed in the San Francisco market place located in commune 3 of the city of Bucaramanga, which currently generates about 16 tons of waste weekly, in which, through the implementation of environmental education strategies To comply with environmental management , we sought to sensitize the population and develop environmental awareness to reduce generation, reuse and recycle, as well as properly separate waste at the source so that it is not taken to final disposal, but to composting. The result was very encouraging, as it was possible to reduce the amount of waste generated by 41.3%, in addition to improving the environment and quality of life of the population where commercial activities are carried out, promoting the environmental sustainability of Plaza San Francisco. .

INTRODUCTION

Until before Law 99 of 1997, solid waste generated in Colombia occupied an insignificant place in the analysis of the priority problems of cities. Not only because the majority of the population thought little about the pollution problems they caused, but also because their disposal was carried out without apparent problems, or at least, without the majority of the population knowing about it. Currently in Colombia, 11,523,807.25 tons of waste are generated per year (Superservicios, 2020) . The composition of this waste is 81% food waste, 13% plastics, 3% paper, 2% metal and 1% glass. According to their origin, they are classified as 73% domestic, 13% commercial, 7% institutional, 4% industrial, 2% construction and demolition and 1% others, according to the Ministry of Environment, Housing and Territorial Development.

Added to the current problem of waste generation and management in the city is the growing global trend of waste production (UNEP, 2002) . In Latin America in the last 30 years, waste generation has doubled, with a regional average of 0.92 kg (UNEP, 2002) . In the case of Colombia, in 2018, 515 kg per inhabitant per year were generated (DANE, 2020) , by 2030 the generation of waste in urban and rural areas could reach 18.74 million tons per year; of which 14.2 million tons of waste per year must be disposed of in landfills that do not have sufficient capacity to receive them, since in the different landfills there would be a deficit in installed capacity, which is estimated at 10.28 million tons by 2030 (DNP, 2016) .

Likewise, the production and final disposal of waste has become one of the main problems worldwide. Globalization, conceived as economic development based on the integration of local economies into a global market economy (Aguilar Benítez, 2011) , in which markets and capital are established on a global scale, along with the definitive establishment of a society determined by hyperconsumerism, which according to Naomi Klein: "We have completely globalized an unsustainable economic model of hyperconsumerism, which is now successfully spreading around the world and is killing us" and the overflowing urbanism that becomes a generation without control of waste. increasing and without the possibility of managing, processing or bringing to final disposal in a controlled and sustainable manner.

Taking into account the above, despite being a relevant aspect, the increase in the acquisition of goods and services in the economy becomes a harmful factor for the environment, since a greater quantity and variety of new products enter every day. , their acquisition and permanent disposal begins. In this way, the model of society has become buy, use and throw away, obtaining more than what we really need, generally without taking into account the effects that this will bring (Franco Antolinez, Meza Joya, & Almeira, 2018) . Likewise, the increase in waste generation and accumulation in landfills has led to collapse, due to the consumption of goods and services, which is a global trend despite the fact that Chapter 21 of Agenda 21 determines the guidelines for the comprehensive management of municipal solid waste as part of sustainable development. It is then determined that solid waste management must contemplate actions to minimize waste generation, recycling, collection, treatment and adequate final disposal. Additionally, it establishes that each country and each city will formulate programs or strategies to achieve adequate waste management according to their local conditions and economic capacity (Acurio, Rossin, Paulo, & Francisco, 2014) .

One option and perhaps the most important to solve the problems associated with waste generation is environmental education, which, according to the EPA, increases citizens' awareness and knowledge of environmental issues or problems. In doing so, it provides the public with the tools necessary to make informed decisions and responsible actions (Thomas-Hope, 1998) . Environmental education is a social tool that allows individuals to achieve significant knowledge of the inhabited environment, reduce the probability of generating environmental impacts and respond to the presence of environmental phenomena to which they are vulnerable (Rojas Hernández, 2007) .

From this same point of view, in 2015 a new global policy was adopted by 193 member countries of the United Nations called the 2030 Agenda for Sustainable Development, which seeks to increase the development of the planet and improve the quality of life for all. human beings with 17 Sustainable Development Goals (SDGs). By 2030, one of the goals is to effectively reduce waste production through prevention, reduction, recycling and reuse. It is increasingly viable to embrace production standards, sustainable consumption and proper management of solid waste generated to significantly reduce damage to the environment and health.

For Colombia it is a challenge, since historically waste management has generated an environmental debt since, for some time now, uncontrolled disposal systems have predominated, within which the deposit of 'garbage' has been included in the bodies of water, generating great water pollution in most of the country's rivers, uncontrolled burning that generates air and soil pollution, open dumps generating visual impact and generation of uncontrolled gases and burials, among others. Within this panorama, market squares are not immune to this problem, hence there is an express need to generate forceful actions to adequately manage the waste generated in these spaces, important for the supply of items that solve basic needs of the community. population.

IDENTIFIED PROBLEM

The inadequate management of municipal solid waste (MSW) generates a growing problem and concern in cities in developing countries. Solid waste generated in cities is usually disposed of in an open dump or landfill, which is not an environmentally viable action for its elimination, since this activity leads to environmental risks that generate ecological imbalances with respect to natural resources. and biodiversity. Municipalities face a big problem with the management and final disposal of the waste generated, taking into account that this entails a very large expense, and is not a necessity for administrations, for this reason, it receives very little interest and investment. It is not only a technical situation, but also a social, environmental, cultural, legal and economic situation, such as, for example, the resources available for the formulation and execution of waste management plans that are structured in such a way that they are not sustainable (Vargas Buitrago, 2016) .

The increase in population and the community's standard of living, rapid economic growth, lack of culture and environmental commitment has led to a very worrying point regarding the environmental problems associated with waste management; which requires an adequate management plan, a lot of will, political and economic commitment, as well as a change of culture in society and a commitment to adequate waste management (Baud, Grafakos, Hordijk, & Post, 2001) . Due to the aforementioned factors, a direct environmental risk to public health is generated in public market squares, affecting the quality of life of the population.

Pollution in the three resources (soil, air and water) is produced by the demand for goods and services from the current forms of production and consumption of anthropic origin that generate a large amount of waste, which together with demographic growth have affected the resilience of strategic ecosystems. In the area of influence where the project is being developed, which is the San Francisco market square, atmospheric and visual pollution is identified, such as the accumulation of critical points, the presence of rodents and buzzards. Additionally, the effects on the final disposal site of this waste generate emissions that pollute the air, leachates that carry toxic substances that contaminate the soil, water and therefore biodiversity. In general, the gases generated contribute to global warming, increasing the crisis due to climate change. From a social and commercial point of view, a negative impact is generated due to the decrease in the quality of life due to public health problems, detriment to its assets, decrease in income due to the impact of its commercial activity, among others.

Can Environmental Education as an environmental management strategy contribute to the reduction in waste generation and its proper management?

CHAPTER DEVELOPMENT

Current Situation of Solid Waste in the World

According to the World Bank, one of the most important byproducts of an urban lifestyle, municipal solid waste (MSW), is growing alarmingly, even faster than the rate of urbanization. Worldwide, ten years ago there were 2.9 billion urban residents generating around 0.64 kg of MSW per person per day (0.68 billion tonnes per year). It is estimated that today these quantities have increased to about 3 billion inhabitants generating 1.2 kg per person per day (1.3 billion tons per year). By 2025, it sets a 70% increase in municipal waste, this will likely increase to 4.3 billion urban residents generating around 1.42 kg/capita/day of municipal solid waste (2.2 billion tons by year). (Review & Management, n.d.) .

Based on the amount of waste generated, its composition and the way it is managed, it is estimated that in 2016 the treatment and disposal of waste generated the equivalent emission of 1.6 billion tons of carbon dioxide, which represents around 5% of total greenhouse gas emissions, while landfills generate 12% of total global methane emissions according to the World Bank's What a Waste 2.0 report in 2018. Additionally, it establishes that, although this is a An issue that people are aware of, the increase in waste generation is at a worrying rate. Countries are developing rapidly, without taking into account efficient technologies to manage the variable composition of waste generated by the population. Cities, where more than half of the people live and where more than 80% of the planet's gross domestic product (GDP) is produced, are in a vanguard position when it comes to facing the challenge of waste at a global level (Kaza, Yao, Bhada-Tata, & Van Woerden, 2018) .

Taking into account that waste production is projected to increase with economic development and demographic growth, it is feasible that lower-middle-income nations will register the highest increase in waste generation. Upper-middle-income and high-income nations provide nearly universal waste collection services, and more than a third of high-income nations' waste is recycled and composted. In low-income nations, about 48% of waste is collected in large cities, only 26% in the rural sector, and only 4% is recycled at the national level. In total, 13.5% of global waste is recycled and 5.5% is composted.

Current situation of solid waste in Colombia

According to the National Planning Department (DNP) with CONPES 3530 of 2008, and after the elimination of inadequate final disposal sites in a successful and verifiable manner, the Comprehensive Solid Waste Management Plans PGIRS acquire relevance as organizational mechanisms of the territories and actions are determined for the development of efficient schemes, as well as regionalization schemes. Additionally, sector strategies are established with the intention of determining a recycling and organized use strategy in towns or regions with technical and economic feasibility for their development. Also, regionalization is sought as a policy strategy for waste management that has lasted over time, established in the different Development Plans at the National level with the purpose of stimulating the implementation of sanitary landfills and regional transfer stations (DNP, 2008) .

In document CONPES 3874 of 2016, with which the most recent National Policy for the Comprehensive Management of Solid Waste was established, the horizon and object of the policy evolves, adding in addition to the health benefit, an environmental, social and economic benefit of such so that it contributes to the promotion of the circular economy, sustainable development, and adaptation and mitigation to climate change. Transformations stand out in which the segregation of waste at the source, the culture of the population, the use of waste and the formalization of recyclers as decisive actors in the collection and transportation of usable waste are of interest. (Superservices, 2020) . On the other hand, according to document CONPES 3874, Colombia will have 64 cities with more than 100,000 inhabitants in 2035, so the generation of waste in urban and rural areas could reach 18.74 million tons annually; of which 14.2 million annual tons of waste must be disposed of in landfills that do not have sufficient capacity to receive them (DNP, 2016) .

Currently, the country generates more than 12 million tons of solid waste per year, of which only 17% is recycled. Of all the waste generated in the country, 96.01% is disposed of in 174 landfills for 973 municipalities of the 1,102 that make up the country. The rest of the waste is disposed of in open-air dumps, temporary cells, and contingency cells. and treatment plants, of which it is worth clarifying that open dumps and temporary cells are not authorized by the environmental authority. Of the 174 landfills that represent 100%, 49.4% have expired their useful life or are about to expire in the short term and the rest have a term of 10 years or more, taking into account that the report is formulated as of December of 2018 (Superservicios, 2020) .

The National Planning Department (DNP) affirms that Colombia in the year 2030 will have health emergencies in most cities and a high generation of greenhouse gas emissions, which affects air quality and contributes to global warming, continue in the same dynamics of waste generation, without finding solutions to improve the reduction and use of this.

Waste situation at regional level

A study carried out by the Mayor's Office of Bucaramanga in 2018 revealed that the city has low environmental quality, and that Bucaramanga is the city that generates the most garbage per inhabitant in the country (Arenas, 2019) . In the environmental impact study developed, it was determined that each inhabitant of Bucaramanga produces more than one kilo of solid waste per day (1,034 kg/inhabitant*day) (SNEIDER, 2018) , even above what the norm says, which speaks of 0.97 kilos per inhabitants per day (DANE, 2020) , this being one of the most important problems, which makes the municipality the largest generator of garbage in Colombia (RESPUESTA_2_SALUDMEDIOAMBIENTE_PROPO_51_2021.pdf, n.d.) .

The Carrasco landfill receives 200,612 tons of waste per year (Superservicios, 2020) , of which the organic material disposed is close to 55%. For more than 20 years, El Carrasco has been the final solid waste disposal site for the metropolitan area of Bucaramanga and 17 other municipalities; ad-portas of a health crisis due to the closure of the landfill, with 5 times more environmental emergencies, the last in September 2017 and currently being declared a public calamity by the municipality of Bucaramanga.

The result of the situation in the landfill is the consequence of administrative, technical, political factors, and by far, is the reflection of the inhabitants as a society, which has not taken responsibility for the serious damage caused by the waste that is generated and the still more serious damage generated by this poorly managed waste. It must be considered that the Carrasco has already completed its useful life and despite this it continued to operate under the declaration of a health emergency and currently under the declaration of a public calamity.

Impacts Generated by Solid Waste

Solid waste constitutes a powerful source of health and environmental pollution; Inadequate management and poor final disposal, in addition to impacting natural resources and the health of populations, directly affects the impact of the landscape. These impacts are direct risks that threaten people's health and generate environmental effects in the place of influence where the waste is generated or disposed of. Indirect risks are constituted by disease vectors and environmental effects that cause landscape deterioration and some type of contamination of natural resources with effects not necessarily in the area of influence of the waste generated and generally in the long term.

Among the potential effects that the inadequate accumulation and disposal of solid waste generates are: fragmentation due to the loss and destruction of the habitat generated when constructing the landfill and its complementary works, such as roads, treatment systems, among others; production of greenhouse gases due to the degradation of the organic matter there, which, in the long term, contributes to climate change and the consequences that it brings; contamination of soil and water due to the change in use and the discharge of leachate respectively with the presence of highly toxic substances, such as endocrine disruptors, of which heavy metals are present in many wastes that are discarded, reach landfills or arrive directly. to the ocean, where they enter the food chain from plankton to humans, which, due to their capacity for bioaccumulation in the long term, contributes to effects on living beings such as the feminization of species, a situation that is of great concern to the scientific community. Also the presence of pests associated with the availability of food and favorable environments for reproduction and the presence of bad odors.

The most serious, but least recognized, environmental effect is the contamination of water, both surface and underground, due to the dumping of garbage into rivers and streams, as well as percolated liquid (leachate), a product of the decomposition of solid waste in open-air dumps. The discharge of solid waste into water currents increases the organic load that decreases dissolved oxygen, increases nutrients that promote the development of algae and leads to eutrophication, causes the death of fish, generates bad odors and deteriorates natural beauty. of this resource.

Another negative impact is soil degradation, caused by the disposal and storage of waste, which generates soil poisoning as a result of the dumping of toxic compounds and disturbs its physicochemical properties. A third negative effect is air pollution, a product of waste deposited in landfills, parks, roads, and streets, which cause respiratory problems and irritation to the eyes and nose, in addition to the discomfort caused by offensive odors. Greenhouse gases from waste are a key contributor to climate change. In 2016, 5% of global emissions came from solid waste management, not including transportation (GONZÁLES, 2016) .

Another environmental problem that Bucaramanga and its metropolitan area currently faces has to do with the presence of vultures “Coragyps atratus”. The buzzard is a scavenger bird whose function in the environment is to 'clean'. However, the overpopulation of this animal has generated multiple difficulties in the metropolitan area of Bucaramanga, which is why an environmental hero has become a villain. According to Gerson Peña D., a biologist from the Bucaramanga Metropolitan Area (AMB), the uncontrolled increase in the population of buzzards in the Santander capital set off the alarms of the environmental authority and Civil Aeronautics.

Landfills and Climate Change

The alarms of the scientific community have been activated by the evident and imminent change in the earth's temperature, which has begun a process of research into the factors or elements that in some way can generate what has been called climate change. Elements that, although in an invisible and imperceptible way, make a great contribution to the production of the so-called GHG or greenhouse gases, precursors of climate change and its consequences. Final disposal sites are means for the accumulation of waste that, due to their relevance within the gas production cycle, require special attention (Bingemer & Crutzen, 1987) .

The generation of biogas in a landfill is the environmental problem that occupies second place in order of importance, after the production of leachate, a product of the accumulation of waste in final disposal sites (Camargo & Vélez, 2009) . Among the impacts attributed to it is its contribution to global warming through the so-called greenhouse effect. The direct and indirect effects of CH4 are calculated to be 20 times greater than an equivalent mass of CO2. The global emission of CH4 is estimated between 20 and 70 tg/year (1 tg = 1012 g) (Bingemer & Crutzen,

1987) . Although this method of final waste storage is extremely active in generating GHG, they are sources that after year five and throughout their useful life, have the possibility of contaminating for many years, particularly with regard to the CH4 methane gas, for which it is determined that it accounts for 13% of total emissions worldwide and where its production will continue even after the landfill is closed, for a projected period of 50 years (Pinzón Uribe, 2013)
.

A landfill acts as a biochemical reactor, with raw materials (waste) and water as inputs, discharges (leachate) and gases as main outputs. Any biodegradable element that is placed in a garbage dump or landfill will receive the action of microorganisms, leading to decomposition and generating gases that are emitted into the air.

In the position of the scientific community that studies the effects of global warming and therefore climate change, it is considered that the greatest number of these impacts will be harmful to the interests of humanity and in tomorrow there will be an "increased greenhouse effect." " generated by increasing concentrations of GHG as a result of the increase in actions developed by man (Common & Stagl, 2008) . The evaluation carried out by the Intergovernmental Panel on Climate Change (IPCC 2000), formed to evaluate the probability of generating warming as a result of the emission of greenhouse gases, in special report number three on emissions scenarios IE-EE determined that greenhouse gas emissions are the result of very complex dynamic systems, predetermined by scenarios such as population growth, socioeconomic development and technological change (White et al., 2001) .

Sustainable Development Goals - SDG

In 2015, 193 member countries of the United Nations adopted a global policy called the 2030 Agenda for Sustainable Development, which contains 17 objectives called Sustainable Development Goals (SDGs). Additionally, the 17 SDGs are integrated, since it is established that interference with one objective will impact the fruits of others and, in addition, development must generate balance in environmental, economic and social sustainability.

In the special scenario of waste, although it is accepted that specific events can positively alter several of the objectives, the indicators to measure the goals against the specific issue focus on only 2 objectives. In SDG 11 "Make cities and human settlements inclusive, safe, resilient and sustainable" and SDG 12 "Ensure sustainable consumption and production patterns".

Target 11.6 proposes, by 2030, to reduce the negative per capita environmental impact of cities, including by paying special attention to air quality and the management of municipal and other waste. Goal 12.5 proposes, between now and 2030, to considerably reduce waste generation through prevention, reduction, recycling and reuse activities.

SDG 12 considers sustainable production and consumption as a goal, with global and local actions as a strategy, to achieve the efficient use of natural resources. This objective also incorporates waste management and the reduction of polluting emissions. Consistent with waste, this SDG can be achieved by minimizing waste production through prevention, reduction, recycling and reuse, both in consumption and production. It is increasingly obvious that the proper management of solid waste and the adoption of sustainable production and consumption models lead to a significant reduction in the negative effects on the environment and health.

In addition, redesign the product life cycle and production chain, properly separate and dispose of waste, take care of food waste and loss, including post-harvest loss. Adopt technologies that reuse waste, make the most of raw materials, think about post-consumption and packaging, linking the producer to the principle of extended responsibility.

Environmental Education as a Waste Management Strategy

An elementary purpose of environmental education is to ensure that both people and communities understand the complicated essence of the environment (the result of the interrelation of its different aspects: biological, physical, cultural, economic, social, etc.) and achieve the values , the knowledge and practical skills to be a responsible and effective solution in the forecast and solution of environmental effects and in the management of environmental quality. From this point of view, environmental education plays a substantial role in facing this challenge, promoting "responsible learning" identified by anticipation and participation that makes it possible not only to understand, but also to get involved in what we must maintain. In this sense, it is necessary to stimulate the appropriation of values, awareness and behaviors that favor the effective and assertive participation of the community in the decision-making process. Environmental education conceived from this point of view can and should be a strategic factor that affects the established development model to reorient it towards sustainability and environmental, social and economic equity.

According to the Ministry of Environment and Sustainable Development 2021, environmental education consists of raising awareness of our reality in the world and our relationship with ourselves and with nature, without forgetting the problems that arise from these relationships. Each of the Sustainable Development Goals (SDGs) fits perfectly with this definition, since they are nothing more than the environmental effects that have arisen from our relationships and the use we are making of nature and that we must solve if we want our future to be be sustainable.

Furthermore, not only can the SDGs be the new basis for environmental education, but this action can be society's best tool to achieve the SDGs. The transversality that environmental education seeks makes it possible to address all these challenges from a single moment with dynamics that make us think and reflect on inequality in different parts of the world, our responsibility towards them and the need to take measures that make us develop. together and better. What environmental education seeks is nothing more than raising awareness of our actions so that an action as simple as using a reusable cloth bag is the first step to ending poverty and protecting the environment. Because that is the basis of environmental education and the SDGs, everything is related and if you want a sustainable world, with all that that implies: equality, sustainable development, an end to hunger and poverty, you must fight to achieve it in every aspect of our life.

Market Square Case Study San Francisco

The food supply centers in the city of Bucaramanga and its metropolitan area generate solid waste with a high use value; However, inadequate management has led to the configuration of critical points due to contamination, deteriorating the urban landscape and endangering public health due to the risk associated with the presence of health vectors. The Carrasco landfill receives 40 tons of waste daily from the market squares of Bucaramanga, according to data from the Bucaramanga Metropolitan Area (AMB). Consequently, they constitute a raw material potential to be exploited. Additionally, the lack of knowledge on the part of all the actors that are part of the productive and commercial chain, added to the lack of direct intervention of the responsible entities, magnifies the problem, tending to reduce the quality of life of users, merchants and inhabitants. of the area of influence and those who use these public spaces.

The Metropolitan Area of Bucaramanga (AMB) as an environmental authority in 2015, formulated resolution 0085 of 2015 based on Resolution 754 of 2014 of the Ministry of

Environment and Sustainable Development, by which it determined the regulations for the formulation, monitoring and evaluation of the Comprehensive Solid Waste Management Plan in market places, called Internal Solid Waste Storage and Presentation Programs. This resolution established the formulation of 8 subprograms as a tool for managing the specific problems identified in each of the market places.

Within the strategies, environmental education is shown as a high-impact alternative to improve the current environmental conditions that were previously described. The solution proposed in the short, medium and long term is of a practical nature, which is structured as a program immersed in the comprehensive solid waste management plan applicable to the market place in accordance with current environmental regulations on the matter. This management instrument involves the development of activities such as awareness-raising, environmental pedagogy, waste characterization, verification of compliance with measures and strategies, control and monitoring. The aim is to join forces between commercial users, public companies and academia to consolidate an efficient mechanism in the management of this solid waste, strengthening a rational culture that is reflected in the minimization of generation, adequate segregation at the source, reuse, recycling, collection, transportation and final disposal.

METHODOLOGY

When identifying the problem in the San Francisco market square, it is necessary to propose actions aimed at minimizing the social, economic and environmental impacts derived from the scenario already described; This is why a management strategy was developed for the market place. This strategy consisted of formulating and implementing the guidelines established by the Bucaramanga Metropolitan Area (AMB). In the development of this strategy, primary importance was given to environmental education to promote a change of culture in the community, translating this into a better quality of life for its population.

The market square is located in the northeastern sector of the urban area of the city of Bucaramanga, with an area of 6,400 m2, in commune 3 of the city. It is a cultural, social and economic welcome center; where various activities are carried out that support commerce in the city. It has 935 stalls distributed among fruits, vegetables, meat, dairy products and others. The overall business process of market places is very practical and constitutes the purchasing, receiving, storing, unpacking, packing and selling of goods.

The methodology developed in the study area that includes the market place presents a quantitative descriptive design, having a main causal relationship where the negative environmental impacts due to the generation of waste constitute the main factor that drives the study, seeking through education environmental a tool for controlling and minimizing the derived impacts, where the final effect is to improve the quality of life of the community in the area of influence of the market square and the Carrasco landfill. Through this methodological development, we seek not only to approach the problem, but also to try to find its causes for its subsequent development and solution through environmental education strategies to transform the consciousness of the community and therefore the environment. These actions were guided by activities structured in the formulated subprograms, in order to generate and evaluate management indices in each of the activities developed.

The methodological framework is the set of steps, techniques and procedures that are used to formulate and solve problems (Fidias G., 2012) . In this sense, to prepare the diagnosis or baseline, surveys, interviews and checklists were carried out. merchant users, administrative personnel, general services personnel, personnel in charge of waste collection and disposal; as well as photographic record of the areas to be evaluated as seen in figure 1.

Figure 1 . *Methodological framework*

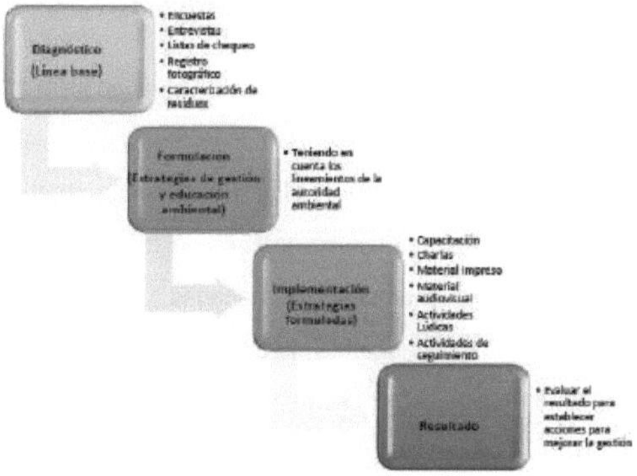

Fountain. Author

Likewise, for the quantification and qualification of the waste, characterization was developed using the statistical sampling analysis method. This method determines that it should be developed for 8 days, taking the information from 7 days, discarding the first, at the time in which the storage unit is operating, separating the waste according to its composition to weigh it later and record the information collected. . This activity allows us to identify the type of waste that is generated in the market place, as well as the amount of waste to establish the actions to be addressed for its proper management.

The project development time frame includes the first phase composed of the diagnosis or baseline and the formulation of the subprograms or environmental education and management strategies that make up the Internal Waste Presentation and Storage program or PGIRS, which took close to 1 year. Once approved by the environmental authority, the second phase contemplated by the implementation of the program began, which was established for a period of 5 years in the short, medium and long term. Currently, the program is in a medium-term execution stage for 3 years. The implementation of the environmental management and education strategies were developed by linking students in the practice modality of the Environmental Engineering and Technology in Environmental Resources program of the Technological Units of Santander, who go daily to the commercial establishment to carry out the stage of implementation of the formulated subprograms, that is, all the activities contemplated for the adequate management of waste, mainly through education, recreational, training, monitoring and control activities. As well as, the evaluation of the results obtained for the improvement of management.

RESULTS AND DISCUSSION

The environmental and health diagnosis makes it possible to identify and interpret the current environmental conditions in which the resources used to provide the cleaning service are found, giving them a qualitative and quantitative assessment of the most relevant environmental effects. The fact that a large amount of organic waste is produced in the commercial establishment, which due to its characteristics manages to decompose quickly, transforms it into an important factor capable of triggering more direct and immediate environmental impacts, which is why it is important to implement strategies to the use of said material that is generated daily.

In the diagnosis or baseline stage, diagnostic tools such as checklists, surveys, interviews, photographic records were applied to identify aspects of waste management, waste presentation habits, separation, recycling, use, knowledge about waste, operators, etc.; with which it was generally identified that there was no knowledge in the plaza on the part of the merchant users about the management of solid waste and in addition there was no established strategy for its management, it was simply available in the room. storage without any handling to be collected by the operator and taken to final disposal. Likewise, the initial characterization of the waste was carried out, which is also part of the diagnosis, in which the characteristics of the waste generated are determined quantitatively and qualitatively, a measurement of the waste presented in the storage room was carried out, during a week, on the plaza's operating day, as seen in Table 1. In the initial characterization, it was evident that no user performs separation at the source or selective storage in the storage room.

Table 1 shows the amount of waste generated weekly in the market place, reaching almost 40 tons per week, with a daily average of 5.5 tons per day.

Table 1. Daily Distribution of Solid Waste Generated

Record of quantitative results (kilograms) of the baseline characterization					
Type of waste generated (kg)	**Organic**	**Inorganic not usable**	**usable inorganic**	**Specials**	**Total day (kg/day)**
Tuesday	3842.62	388.04	352.9	986.83	5570.39
Wednesday	4954.9	397.98	218.84	883.78	6455.5
Thursday	3630	218.97	159	1037.86	5045.83
Friday	3643	435.25	218.88	1429.77	5726.9
Saturday	3671	336.7	155.35	841.5	5004.55
Sunday	3476	200.28	111.31	844.2	4631.79
Monday	4915	343.96	112.71	1309.05	6680.72
Total week	28132.52	2321.18	1328.99	7332.99	39115.68
% in the composition	71.9	5.9	3.4	18.7	100

Fountain. Author

Figure 2 shows that 71.9% of the waste generated in the San Francisco market square is organic waste, usable inorganic waste represents 3.4%, non-usable inorganic waste represents 5.9%, and non-usable inorganic waste represents 5.9%. specials 18.7%.

Figure 2 . Percentage distribution of waste generated

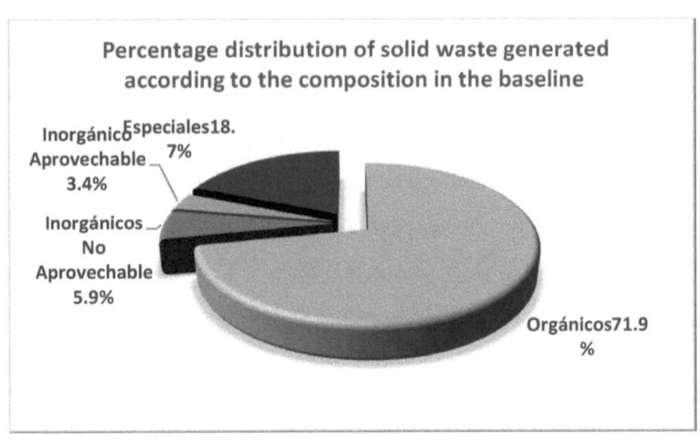

Source: Author

Once the diagnosis, formulation and approval stage of the programs was completed, the implementation stage of the proposed awareness-raising, education, monitoring and control activities began. Activities such as training, talks, public speaking and dissemination of pedagogical material on important aspects to take into account in waste management; Likewise, inspection of the storage room, waste collection and transportation vehicles and monitoring of the separation at the source of waste by each successful bidder. All activities are important, but special monitoring is carried out on separation at the source, since it is important that all successful bidders deliver waste separated according to its composition to carry out selective collection and use of organic plant waste through composting. . Likewise, environmental education activities, to generate a change in behavior that leads to social transformation.

Table 2. Daily distribution of solid waste generated after 3 years

Record of quantitative results (kilograms) of the characterization 3 years later					
Type of waste generated	**Solid waste Organic**	**Unusable inorganic solid waste**	**Reusable inorganic solid waste**	**Special solid waste**	**Total, solid waste**
Tuesday	2387.07	169.8	68.65	59.65	2685.17
Wednesday	1720.92	173.62	15.05	120.8	2030.39
Thursday	1654.9	89.5	1.05	68.06	1813.51
Friday	2148.72	128.3	41.55	30.7	2349.27
Saturday	2154,166	187.85	22.9	136.08	2500,996
Sunday	1766.54	45.62	19.55	21.75	1853.46
Monday	2815.49	63.82	29.23	22.4	2930.94
Total (kg)	**14647,806**	**858.51**	**197.98**	**459.44**	**16163,736**
% in the composition	**90.61**	**5.31**	**1.22**	**2.86**	**100**

Source: Author

Table 2 shows the amount of waste generated weekly in the market place after 3 years of implementation of the activities contemplated in the programs, generating about 16 tons weekly, which compared to the baseline capacity (40 tons approximately) corresponds to 41.32% of the waste generated at that time.

Figure 3 shows that 90.61% of the waste generated in the San Francisco market square is organic waste, usable inorganic waste, non-usable inorganic waste and special waste represent close to 10%. Compared with the characterization of the baseline, it can be stated that apart from the fact that waste generation has been minimized by more than 50%, the separation process at the source has also been improved through the generation of environmental awareness and culture, In this way, it has been possible to take advantage of a large percentage of the waste generated in the square.

Figure 3. Percentage distribution of waste generated

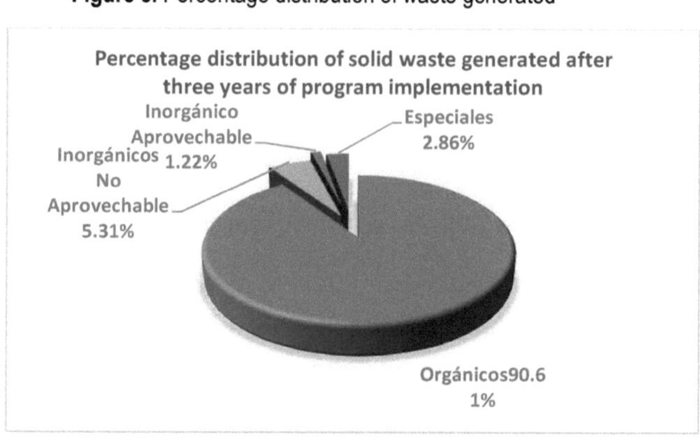

Source: Author

CONTRIBUTION TO THE SDGS AND SUSTAINABILITY

The 2030 agenda requires modifying the current production and consumption model towards a circular economy, this in effect determines doing more with less, with fewer resources, with less energy expenditure and generating less waste. The 2030 agenda proposes in SDG 12 to guarantee sustainable consumption and production patterns, in which there is an express goal, 12.5, which in its statement proposes to considerably reduce the generation of waste through prevention, reduction, recycling and reuse activities. here to 2030, seeking to control the excessive exploitation of natural resources and change the current production and consumption model towards a circular economy.

Good comprehensive waste management involves a series of interrelated actions that allow the environmental, health, economic and social situation in a territory to be modified over time. This management must be determined by environmental education actions, community awareness campaigns, separation at the source, prevention activities, reuse and recycling, use or treatment according to the characteristics of the waste and a controlled final disposal. Therefore, it is important to change the production model and consumption habits and encourage both strategies through education and generation of environmental awareness in the communities.

Work with the community must be the backbone for the development of the plan, since only the response and community actions will determine its success or failure. The first stages of the Environmental Education and Awareness Campaign, together with educational institutions and official entities, are essential for the community to respond to waste management strategies such as separation at the source in each workplace, reuse and recycling. Within this process, recyclers play an important role to recover waste that can be converted into raw materials and add value to it, to return it to a productive cycle in a circular economy.

From that point of view, which SDGs are you contributing to?

It could be said that several of these objectives are being contributed directly or indirectly with the formulation and implementation of the Program in the market place. Taking into account that the SDGs are interdependent since, if actions are developed in one objective, they will be reflected in another, from this point of view it can be said that a smaller contribution is made to the SDGs: 1. End of poverty; 2. Zero Hunger; and 8. Decent Work, referring specifically to the impact that would be generated by being able to help current recyclers in their unworthy socioeconomic situation and thus be involved in the productive cycle. From the environmental aspect, it impacts: 6. Clean water and sanitation. In greater proportion in the SDGs: 11. Make cities and human settlements inclusive, safe, resilient and sustainable; 12. Sustainable Production and Consumption; and 13. Climate Action, given that it is one of the factors that contribute to the generation of CO2, CH4, in short, Greenhouse Gases that lead to Climate Change.

Specifically for SDG 11 and SDG 12, it had previously been indicated that the specifically proposed goals on waste indicated that: Goal 11.6 proposes, between now and 2030, to reduce the negative per capita environmental impact of cities, including paying special attention to the air quality and municipal and other waste management. Goal 12.5 proposes, between now and 2030, to considerably reduce waste generation through prevention, reduction, recycling and reuse activities. From this point of view the contribution is complete, reducing the impact on air quality by reducing the generation of greenhouse gases and establishing, through environmental education strategies, actions for prevention, reduction, recycling and reuse with evident results in the specific case of the San Francisco market square, improving the quality of life of the population in the area of influence of the project.

DISCUSSION AND CONCLUSIONS

Despite all the efforts initiated worldwide in terms of recycling and responsible use of resources, the issues of municipal waste management, together with climate change, will gain weight worldwide. Correctly managing waste is essential for the construction of sustainable and liveable cities, and continues to be a challenge for many developing countries and cities, such as ours.

In the diagnosis stage, the aim was to identify aspects and information from commercial users that would lead to determining environmental education processes, such as awareness, training, changes in behavior and waste generation habits, with which it was concluded that no no environmental education process had been developed, nor was any waste management process carried out; This is how environmental education is established as the main strategy, to improve waste management in the square and reduce impacts. With the intervention in the square mainly through environmental education tools, it has been possible to demonstrate the importance of raising awareness and creating environmental culture to improve environmental quality and therefore the quality of life of the community. Before starting the awareness and education process in the market place, no action was taken for adequate waste management, that is, it was not recycled, reused, or separated at the source, much less used; The use of special waste generated in the plaza was barely managed. Currently, the process has managed to separate at the source, properly present the waste, reuse and recycle, to the point that, compared to the characterization of the baseline, it can be stated that, apart from the fact that the generation of waste has been minimized by more than 50%, the separation process at the source has also been improved through the generation of environmental awareness and culture, in this way it has been possible to take advantage of the majority of organic waste generated in the square, which constitutes 90.6%.

Once the diagnosis and intervention of the plaza was carried out through pedagogical, inspection, monitoring and control tools, it was possible to demonstrate the importance of the results of these strategies. The decrease in direct environmental impacts is evident in the decrease in critical points due to inadequate waste management, the decrease in the discharge of leachate mainly in the storage room, therefore the decrease in unpleasant odors, the decrease in pests, rodents, insects, buzzards, etc. And the improvement in the image of the square that, as an added value, brings an increase in economic productivity. Regarding indirect impacts, the decrease in the amount of waste that is taken to the landfill should definitely reduce the impacts generated on environmental resources in the area of influence of the landfill. It is important to continue with the process in search of continuous improvement, managing to fully manage the waste generated in the plaza; which will mainly seek to create an environmental culture of all the actors that are part of the production chain.

RECOMMENDATIONS

The organic solid waste of plant origin generated in the San Francisco de Bucaramanga market place constitutes a potential raw material (approximately 16 tons/week) to be subjected to a biological and chemical transformation and valorization process that would generate economic and social benefits. and environmental, given that its use would allow the valorization of all waste that has become a problem due to the difficulty of its management and disposal, without taking into account the high costs that this entails. This alternative should serve as a basis for the other sectors that generate this type of waste to take into account, mainly the market places of the Metropolitan Area, which produce about 220 tons/week of solid waste of organic origin, which would not be disposed of in the Carrasco landfill.

The advances, both in the gradual acceptance of the standards, as well as in the reduction, reuse, recycling, separation at the source by citizens and separate collection by the companies providing the collection service, show the value of awareness to mobilize inertia, create an environmental culture of resistant postures to collaborative postures. This is possible if differentiated environmental education strategies are used, depending on the target group to intervene, as was the case of the merchant users of the San Francisco market, which demanded and demands knowledge of the characteristics and evaluations of each group; That is, knowing the perceptions, appreciations, context and possible resistance that a group presents to what it wants to promote and transform.

Finally, one of the main recommendations that emerge from this analysis is to consider the role that the government must assume as a promoter of change in society; a role that can lead to a gradual transformation of the impression that the community has of its government and that can stimulate an argument that leads to the solutions that any government must seek for the environmental problems it addresses; For this, projects, programs, and policies require knowing and recognizing the appreciation of the community, the important role it plays, having consistency and permanence in the development of the proposals, evaluating the result to generate changes if necessary, beyond administrative changes and report periodically and frequently on the achievements, benefits and difficulties with which the institutions advance their goals, in this way progress is made on the path of sustainability.

BIBLIOGRAPHIC REFERENCES

Acurio, G., Rossin, A., Paulo, T., & Francisco, Z. (2014). Diagnosis of the situation of the municipal solid waste collection in Latin America and the Caribbean. *Croquis* , *2* (34), 130.

Aguilar Benítez, O. (2011). The globalization process and the current capitalist financial crisis. *Working Papers* , *67* , 1–18.

Arenas, JLP (2019). Bucaramanga will insist on the transformation of garbage. *Vanguard* , 7–8. Retrieved from https://www.vanguardia.com/area-metropolitana/bucaramanga/bucaramanga-invertira-en-la-transformacion-de-las-basuras-gy1652190

Baud, I., Grafakos, S., Hordijk, M., & Post, J. (2001). Quality of life and alliances in solid waste management. Contributions to Urban sustainable development. *Cities* , *18* (1), 3–12. https://doi.org/10.1016/S0264-2751(00)00049-4

Bingemer, H.G., & Crutzen, P.J. (1987). The production of methane from solid wastes. *Journal of Geophysical Research* , *92* (D2), 2181–2187. https://doi.org/10.1029/JD092iD02p02181

Camargo, Y., & Vélez, A. (2009). Biogas emissions produced in landfills. *II Ibero-American Symposium on Waste Engineering* , (September 2009), 1–12.

https://doi.org/10.1128/JCM.39.2.560-563.2001

Common, M., & Stagl, S. (2008). Introduction to ecological economics. In *Introduction to Ecological Economics* .

DANE. (2020). Technical Bulletin Environmental and economic account of material flows – solid waste. *Dane* , 1–19. Retrieved from https://www.dane.gov.co/files/investigaciones/pib/ambientales/cuentas_ambientales/cuentas-residuos/Bt-Cuenta-residuos-2018p.pdf

DNP. (2008). Guidelines and strategies to strengthen the public sanitation service within the framework of comprehensive solid waste management. *National Council of Economic and Social Policies of the Republic of Colombia* , *53* (9), 1689–1699. Retrieved from https://colaboracion.dnp.gov.co/CDT/Conpes/Económicos/3530.pdf

DNP. (2016). CONPES Document 3874. National Policy for the Comprehensive Management of Solid Waste. *National Council of Economic and Social Policy Republic of Colombia. National Planning Department (DNP)* , 1–73. Retrieved from https://colaboracion.dnp.gov.co/CDT/Conpes/Económicos/3874.pdf

Phidias G., A. (2012). The Research Project. In *Journal of Chemical Information and Modeling* (Vol. 53). Retrieved from https://drive.google.com/file/d/0ByOr72_-tQvdWkpyNG9URmNPWGh1ZWlsTkpndlVCT0ZQNjdn/view?pli=1

Franco Antolinez, LJ, Meza Joya, MA, & Almeira, JE (2018). Situation of the final disposal of solid waste in the Metropolitan Area of Bucaramanga: El Carrasco landfill case (review). *ADVANCES: Engineering Research* , *15* (1), 180–193. https://doi.org/10.18041/1794-4953/avances.1.4735

GONZÁLES, JA (2016). Solid waste: problem, basic concepts and some solution strategies. *Management and Region Magazine* , (22), 101–119. Retrieved from https://revistas.ucp.edu.co/index.php/gestionyregion/article/download/149/146

Jaramillo, J. (1999). International Seminar: Comprehensive management of solid and hazardous waste, 21st century. *Comprehensive Management of Municipal Solid Waste - GIRSM* , 1–20.

Kaza, S., Yao, L.C., Bhada-Tata, P., & Van Woerden, F. (2018). *What a Waste 2.0 : A Global Snapshot of Solid Waste Management to 2050* . Retrieved from http://hdl.handle.net/10986/30317

Pinzón Uribe, LF (2013). *Influence of Landfills on Climate Change* . (March), 1–13. Retrieved from http://www.umng.edu.co/documents/10162/745277/V2N1_8.pdf

Answer_2_Saludmedioambiente_Propo_ *51_2021.pdf* . (n.d.).

Review, G., & Management, S.W. (n.d.). *A Global Review of Solid Waste Management* .

Rojas Hernández, J. (2007). Human Community, Development and Biosphere. Towards Comprehensive Sustainability. *Journal of the Unesco Chair on Sustainable Development* , *1* , 9–27. Retrieved from http://www.ehu.eus/cdsea/web/wp-content/uploads/2016/12/Revista1.pdf#page=31

SNEIDER, F. P. (2018). *SIMULATION MODEL OF THE ENVIRONMENTAL IMPACT OF WASTE IN BUCARAMANGA AND ITS METROPOLITAN AREA* . AUTONOMOUS UNIVERSITY OF BUCARAMANGA.

Super services. (2020). National Report on Final Disposal of Solid Waste. *Superintendence of Home Public Services* , 142.

Thomas-Hope, E. M. (1998). *Solid waste management: critical issues for developing countries* (1st ed.; C. Press, Ed.). Retrieved from https://books.google.com.co/books?hl=es&lr=&id=QACfDLFXzncC&oi=fnd&pg=PR7&dq=ENVIRONMENTAL+EDUCATION+for+waste+management&ots=ZdANB44Mof&sig=wL4Um2vz-K1dzykONC-gxX_ChFU&redir_esc=y#v= onepage&q=ENVIRONMENTAL EDUCATION for waste management&f=false

UNEP. (2002). Global Environment Outlook 3. In *United Nations Environment Programme* . Retrieved from http://www.unep.org/geo/GEO3/english/pdf.htm

Vargas Buitrago, A.J. (2016). *Technical Study for the Valorization of Solid Vegetable Waste from the San Francisco Market Square in Bucaramanga. (Case study)* . Bucaramanga.

White, KS, Ahmad, QK, Anisimov, O., Arnell, N., Brown, S., Campos, M., ... Wratt, D. (2001). Technical Summary in Climate Change 2001: Impacts, Adaptation and Vulnerability. *Europe* , 19–73.

CAPITULO 2.
DETERMINACIÓN DE DIFERENTES MÉTODOS PARA LA ELABORACIÓN DE CONTENEDORES BIODEGRADABLES PARA EL ALMACENAMIENTO DE COMIDA

EPISODE 2.

DETERMINATION OF DIFFERENT METHODS FOR THE PREPARATION OF BIODEGRADABLE CONTAINERS FOR FOOD STORAGE

Natalia Alexandra Bohórquez Toledo

Chemistry, M.Sc. in chemistry

Research professor of the GRIIV research group, assigned to the Environmental Engineering program of the technological units of Santander

nbohorquez@correo.uts.edu.co

Carmen Lorena Marín Cordero

Environmental Resources Technologist

Student researchers from the GRIIV research group, assigned to the Environmental Engineering program of the technological units of Santander

Lizmalore3@gmail.com

Marly Yulitza Suárez Torres

Technologists in Environmental Resources

Student researchers from the GRIIV research group, assigned to the Environmental Engineering program of the technological units of Santander

marlysuto0427@hotmail.com

ODS:

SUMMARY

The objective of this chapter was to determine the different methods for the manufacture of biodegradable containers for food storage made with materials of natural origin that can be found in the department of Santander, in order to find the materials with the greatest potential for replace conventional products. For this, 3 stages were carried out, the first in which the various materials of natural origin that could be used to manufacture food containers were identified, for this a bibliographic review of different experiences related to the topic was carried out and each experience was analyzed. described the objective, the methodology used, the results obtained with its discussion and finally the best conditions for manufacturing and the properties obtained for that material. Finally, a summary table was made to review the properties of the final products and a comparative table of these properties to recognize which of the authors obtained a good material with respect to the properties of plastics such as polystyrene and polypropylene. The annual production of the various raw materials was also reviewed to check their availability. After all the analysis, it was concluded that a viable proposal would be the manufacture of a food container with a pineapple crown and an internal layer of a material composed of corn starch and bovine gelatin.

INTRODUCTION

The environment is constantly affected by various factors, among the pollution that is generated, is that caused by the use of plastics such as expanded polystyrene (PET), among others, which are used as food containers and are used only once. and they are discarded.

According to reported data, utensils such as disposable plates take between 1,000 or 200,000 years to degrade (Elnoticom, 2018) and the polystyrene used in the well-known disposables, glasses and plates takes 50 years to degrade (Edacción Vida - ELTIEMPO, 2019). Contamination occurs largely because materials such as expanded polystyrene do not have a recycling method (García, cited in BBC News, 2015). According to reports from the United Nations, it is considered that 80% of all the garbage present in the ocean is plastic (UN, cited in Editorial Vida – ELTIEMPO, 2019), and that around 8 million tons of plastic they reach the sea every year (Elnoticom, 2018).

Since the problem is so serious and seeing the negative impact that these utensils generate on ecosystems, their use has been prohibited in many places in the world, including countries such as Jamaica, Grenada, Barbados, Dominica, Belize, Trinidad and Tobago, where from In 2020, single-use plastics were prohibited, including expanded polystyrene (Morales, 2019). In Colombia, in the department of Santander, there is also concern about the problem that these materials cause and that is why in municipalities like La Belleza, a ban was decreed on the use of single-use plastics and expanded polystyrene in all contracts that the municipality makes. with different entities (Ríos, 2020). The Government of Santander signed decree 164 of 2020, in which the same prohibition is made regarding single-use plastics and polystyrene. This prohibition seeks to encourage small and medium-sized companies to produce biodegradable containers (Government de Santander, 2020).

As these utensils represent a large-scale problem, many researchers have tried to find materials that replace plastics, so that these developed materials are environmentally friendly. With the industrial potential and the environmental problem that expanded polystyrene and other plastics represent, we sought to determine and present various investigations in which materials were manufactured that could replace the current ones to produce food containers, and that were manufactured with raw materials from natural origin present in the department of Santander, and thus propose the development of a container that meets the needs of the market, which would help reduce environmental pollution in the future.

IDENTIFIED PROBLEM

For some time now, plastics have been an environmental issue due to the implications they have on environmental pollution. Among the plastic products that pollute the planet are products known as single-use products, among which are food containers such as disposable cups, plates and cutlery. These products are mostly made of polystyrene, a plastic that, apart from not being able to be recycled, takes many years to degrade (Avalos and Torres, 2018).

Most of the plastics used today are produced with raw materials of petrochemical origin, and although recycling these products is one of the options to reduce their environmental impact, this does not completely eliminate the problem, because there are cases such as expanded polystyrene, used to manufacture glasses, cups, plates and food containers, which cannot be recycled due to their cost and the difficulty in collecting, moving and treating them (García, 2015).

Plastics represent such a big problem that there has been talk throughout the world about the island of garbage present in the Pacific Ocean, between Hawaiian territory and North America, in various maritime currents of great importance and which is estimated to have a size of Australia and is made up of various plastics in terms of type and size (Castell6n, sf)

The problem with these products is that they are widely used and it is profitable for the industry to continue manufacturing them in the traditional way, since economics plays an important role when manufacturing a product. Another factor that aggravates the problem is the increase in demand for the use of these products, as the population grows, consumption grows and the lack of environmental education does not allow citizens to realize the serious damage they cause to ecosystems. with the continuous use of these products (Salgado and Granja, 2016).

Demand, more than demand, is a need of the community and continuing to produce products of this style will only generate an increase in pollution, since, even if there is environmental education, the need will always be present. For this reason, as agents of change, there is a need to publicize the advances that the scientific community has had to solve this problem and which of these solutions is the most feasible to make a change to conventional products.

The world population will continue to increase, and the demand for products such as glasses, plates and trays to transport food will also increase, so it is urgent to find a solution that helps the environment without affecting the needs of the population. . For this reason, it is very convenient at this stage of scientific progress regarding the subject, which of all the possible solutions is the one that presents the greatest benefits or whether several of them are able to provide a service equal to that of the conventional plastics that are used today in day to make the mentioned products.

Knowing how they are manufactured, their properties and identifying which ones have the best performance for mass use, allows us to focus all efforts on achieving, in some time not too distant, a production of these products that are economically viable, and that do not pollute the environment. Furthermore, with the knowledge of the advances on the subject, the academic and scientific community contributes to society in terms of sustainable development. Knowing, for example, that new alternatives cannot necessarily be recycled, but they do take much less time in their degradation process in the soil and that the degradation components will not affect it (Salgado and Granja, 2016).

RESEARCH METHODOLOGY

The work was developed in three stages, below, the methodology used in each stage is described and after this description are the results obtained.

Stage 1: Identification of possible materials of natural origin available in the department of Santander that meet the properties for the manufacture of biodegradable containers for food storage

To identify possible biodegradable materials that meet the aforementioned conditions, a bibliographic review of different sources was carried out. To select the sources that fit the objectives of the work, the flow chart in Figure 1 was followed. Once the material was selected, the objectives, the methodology used, the results obtained, and the discussion of said results were described. and the best material and the best conditions to manufacture it.

Figure 1. Flow diagram for selecting research related to food containers made with products of natural origin.

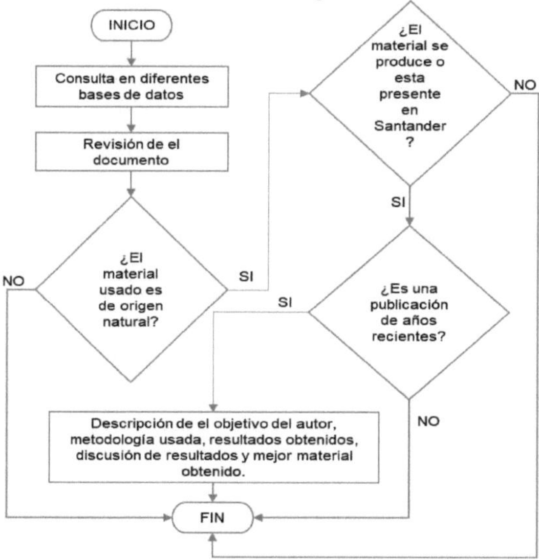

Fountain. Author.

Once all the information was collected, a summary table was made which highlighted the properties of the material chosen by each of the authors as the best material, the manufacturing method of the material, the author of the research and the main raw materials. used for the manufacture of each material.

Stage 2: Annual production of the various raw materials

A comparison was made of the annual production of different raw materials in the department of Santander, to see the availability of material to generate a product. This was carried out through a bibliographic analysis that allowed us to identify the annual quantities of raw materials generated in the department and their availability.

Stage 3: Proposal for the manufacture of a biodegradable container for food storage based on the information analyzed.

With the analyzed properties and the availability of the product, the final proposal was made, taking into account which of these raw materials are produced in the department in greater proportion and carrying out an analysis regarding the non-affectation of food safety.

RESULTS AND DISCUSSION

Stage 1: Identification of possible materials of natural origin available in the department of Santander that meet the properties for the manufacture of biodegradable containers for food storage.

When carrying out the bibliographic review in the different academic search sources, a large number of works were found in which various materials were used, however, only those in which they worked with raw materials that can be obtained in the market were taken into account. department of Santander. Bibliographic references were chosen that talk about the possible materials with which the current expanded polystyrene and other materials of plastic origin that are used to contain food can be replaced. Table 1 shows a summary of the properties found by the authors and the methods used to manufacture the most optimal materials. The authors who did not report quantitative properties, but rather qualitative properties, were marked with green in the summary table.

Table 1.
Summary of properties and manufacturing processes of different materials made with products of natural origin

Main materials	Author(s)	Properties found	Fabrication process
Cooked banana leaf	Iturbe, 2016	Rough texture, good rigidity, weight-bearing	Layering
Banana leaf, carboxymethylcellulose and banana leaf fiber interlayer	Mayorga, Rodríguez and Martínez 2018.	No water leaks through the container, it does not deform, as long as it is treated well and it supports three boxes with 500g of food.	Pressing and die cutting
banana bandana	Corbella, Hernández and Laz, 2018	Tensile strength of 8.17 N/mm^2 Elongation percentage of 2.98%. Young's modulus of 3.35 N/ mm^2. Flexural strength of 1,005 MPa. Maximum load of 12.02N. Bending strain of 21.45% Flexural modulus of 23.40MPa	Pressing
Rachis, pseudostem and banana peel	Pérez and Ramos, 2019	Humidity: 11 -11.6% Ash: 1 -6% pH: 7.84 – 8.74 Weight: 50.66 – 98.76 g/m^2 Density: 10132 – 19753 g/m^3 Thickness: 0.138 – 0.42mm	Pressing
Dried banana leaf and cassava starch	López and Fajardo, 2019	Good utility and easy storage	Pressing
Oxidized banana starch	García et al, 2011	Humidity: 9.04% Ash: 0.29% Total starch 87.17% Apparent amylose 14.44% Viscosity: 3113 UB Glass transition 63.6°C	Casting
Oxidized banana starch and glycerol	Canaza, 2012	More density with more glycerol and less starch. Thicker when glycerol is added Elongation: 21.28% and 27.67% Thickness: 46.61µm and 49.86µm Density: 2.36g/ $m^3$and 2.85g/m^3	Casting
banana flour	Ortega, 2016	Starch: 76.21% Amylose: 15.22 Fiber: 3.18% Deformation resistance of 25.47N 12.68mm elasticity Water permeability of 6.89 E-10g/m*s*Pa Solubility: 60.02%	Casting
Banana peel and cornstarch	Pizá et al, 2017	Force applied to break: low Time it takes to break: 17 seconds	Casting

Banana peel starch	Bejarano, 2018	Maximum stress: 0.91MPa and 1.20MPa. Breaking stress: 0.82MPa and 0.24MPa Young's modulus: 0.09 and 0.37 GPa Maximum Load: 138.67 and 249.84Kgf Elongation: 70.74 and 20.88mm Density 1.15 and 1.01 g/cm^3	Casting and baking
Banana peel starch and chicken feet collagen.	Carvajal, 2019	Elongation percentage: 85.71% Area: 42.35cm^2 Breaking strength:5.42$^g/_{cm^2}$	Casting
corn cob	Puello and Zabaleta, 2014	Humidity: 31.2% Ash: 97.84% Nice smell Scent: wood Texture: rough	Casting and Laminating
Cornstarch	García, 2015.	Tensile strength: 0.25MPa Thickness: 0.51mm Good biodegradability in soil after 3 months	Casting
Corn starch, bovine gelatin and antimicrobial	Diaz, 2015	Thickness: 0.061mm Modulus of elasticity: 512MPa Breaking stress: 24MPa Elongation: 35% Water vapor permeability: 5 EXP 7 g/Pa*s*m Oxygen permeability: 2.08 EXP 13cm^3 /(m^2 * $día$ Equilibrium humidity: 8.4 g water/g dry film Degradation start temperature: 123.5°C	Thermomolding or casting
Corn starch, glycerol and pH variation.	Narvaez, 2016	Modulus of elasticity: 0.59MPa Ultimate Effort: 2.69 Final elongation: 0.109	Casting
Cornstarch	García et al, 2019	Low resistance with a break time of 23 seconds Puncture resistance of 24N for 24 seconds BOD/COD = 0.5	Extrusion
Corn, cassava and glycerol	Muñoz, 2014	Thickness: 0.30mm Water vapor permeability: 1.92 g*mm/h* m^2*KPa Solubility in water: 57.04% Color difference:30.64 Opacity: 3.32 Drilling force: 5.24N Drilling deformation: 7.82% Stress: 1.47MPa Elongation: 54.93% Modulus of elasticity: 0.08	Casting
Corn, cassava and passion fruit fiber	Chariguaman, 2015	Thickness: 0.19mm Solubility 39.88% Opacity: 1.48nm Drilling deformation: 38.98% Elongation strain: 11.7% Modulus of elasticity: 0.74MPa	Casting
Cassava and coffee cisco	Cardenas and Medina, 2011	Good viscosity Ease of handling Good toughness Ease of demolding Fracture strength: 12.72Mpa Modulus of elasticity: 1034.35MPa Elongation percentage: 3.27%	Vacuum molded and baked

Cassava starch and glycerin	Velasco et al, 2012	Internal morphology without phase separation	Extrusion
Yucca and aloe vera	Pretell, 2018	Good flavor, aroma, resistant and flexible. Total ash: 0.7g/100g sample	Casting and baking
Cassava starch and fique fiber	Luna, Velasco and Villada, 2009.	Effort: 11.93 N/cm^2 Elongation percentage: 2%	Casting
	Navia, 2011	Tensile elastic modulus: 366.66MPa. Tensile strength of 1.75MPa. Flexural strength: 3.5MPa Elastic modulus of elongation: 380.51MPa Impact resistance: 17 J/m	thermocompression
	Ayala and Villada, 2015	Tensile breaking stress: 3.2 MPa Tensile elastic modulus: 784.7MPa Bending breaking stress: 8.8MPa Bending elastic modulus: 791.6MPa	Thermocompression
	Navia, Ayala and Villada, 2015	Density: 0.40g/cm^3 Flexural strength: 8.8MPa	Thermocompression
Potato starch	Charro, 2015	Humidity: 9.82% Amylose: 23.52% Thickness: 0.33mm Solubility in water: 25.99% Permeability: 0.0946 g/h*m*MPa Breaking strength: 7.68$^{Kg}/_{cm^2}$ Elongation at break: 18.34$^{Kg}/_{cm^2}$	Casting and warm-up
Potato starch and glycerol	Batuani 2015	Good malleability, flexibility, elasticity and resistance to stress. Its hardness is acceptable.	Casting
Potato starch with chitosan and xanthan	Alarcón and Arroyo, 2016	Thickness: 0.13mm Maximum pulling force: 8.47N Elongation percentage: 33% Elongation/traction ratio: 3.9	Casting
Potato peel starch	Meza, 2016	Maximum stress: 1,470MPa Maximum elongation: 16.822% Biodegradability of 64.21% at 92 days	Casting
Potato starch, polyvinyl alcohol and glycerin.	Holguin, 2019	Elongation percentage: 4.132% Tensile strength: 4.12MPa D shore hardness: 71.2	Reinforced matrix
Arracacha starch	Pardo and Velasco 2011	Tensile strength: 10.485MPa. Elongation percentage: 15.665	Casting
Arracacha, goose and sweet potato starch	Espina, Cruz-Tirado and Siche, 2016	Density: 0.540$^{g}/_{cm^3}$ Grammage:0.135$^{g}/_{cm^2}$ Color difference:13 Hardness:3.62g Fractureability: 2.22mm Flexural strength: 0.16-0.21$^{N}/_{mm^2}$	Thermocompression

Arracacha, goose and sweet potato starch		Density: 0.571$^{g}/_{cm^3}$ Grammage:0.143$^{g}/_{cm^2}$ Color difference:8.27 Hardness:3.99g Fractureability: 1.62mm Flexural strength:0.135-0.175$^{N}/_{mm^2}$	
coconut tow	Quintanilla, 2010	High cellulose content Good rigidity and toughness Impact resistant Low heat conductivity Odorless and moisture resistant	Molded
Coconut tow and rice gum	Cruz et al, 2018	Good toughness Nice color	Molded and pressed
polylactic acid	Tejada et al, 2007	Tensile strength: 40-60 MPa Tensile Modulus: 3-4GPa Maximum temperature: 50-60°C	Fermentation and polymerization
Polylactic acid and polycaprolactone	Valdés, Balart and Reig, 2015.	Elastic tensile module. 828.21MPa Tensile strength: 38.95MPa Elongation percentage: 112.6%. Young's modulus: 2327MPa Bending elastic modulus: 2047.1MPa Flexural strength: 64.8MPa D shore hardness: 63 Charpy Impact: 0.268 Degradation start: 280°C Glass transition: 60-62°C	Extrusion
Sapote and teak leaves	Chacón, 2019	Thickness: 270 μm Density: 0.52v$^{g}/_{cm^3}$ Weight: 154.07$^{g}/_{cm^2}$ Water absorption: 4.6$^{g}/_{cm^2}$ Grease resistance: 3.3$^{g}/_{cm^2}$ Blade stiffness: 0.11g	Pressing
rice husk	Avalos and Torres, 2018	Good toughness Nice color	Pressed and baked
pineapple crown	Kroefly, Robles and Varga, 2018	Good physical properties	Molded
Guadua fiber and Orellana mycelium	Atencio et al, 2018.	Maximum load: 4682.41N Fracture energy: 0.207J	Natural growth in mold
Residual sugarcane biomass	López and Largo, 2018	Humidity: 8.1%	Thermoforming
Orange peel	Giraldo et al, 2018	Maximum compression: 65Kg Impact: 0.530lb/ft Shore hardness: 90.25	Molded and baked

Jelly	Ranaldi, Roldán and Urdinola, sf	Maximum service time: 30 minutes with content temperatures at 13°C, disintegrates quickly at 48°C. Loses original shape over time	Casting
Soy protein and montmorillonite	Echeverria, 2012	Thickness: 10µm Opacity: 89 AU/mm Luminosity: 1.2 Water content: 84.4% Solubility: 17% Water vapor permeability: 3.3 g/s*m*Pa Breaking stress: 8.9MPa Deformation at break: 8.6% Elastic modulus: 5.2MPa Glass transition: -40.7°C	Heating molding
Amaranth seed starch	González and Tovar, 2018.	Humidity: 13.61% Ash: 0.07% %amylose/%amylopectin: 24.15/75.85 Density: $1.25 ^{g}/_{cm^3}$ D shore hardness: 33.10 Maximum effort:2.04 Elasticity:2.24MPa	Starch extraction with NaOH
Wheat gluten	Zarate, 2011	pH =9 Good mechanical and thermal properties	Extrusion or thermomolding
Citrus waste	Arévalo et al, 2010.	Tensile strength: 5.89 – 11.29 MPa. Elongation percentage: 3.02% - 5.9% Thickness: 0.09 – 0.16mm Water vapor permeability: 1.61EXP -5 to 5.65EXP -5 g/ $h.mm^2$	Casting
Mango seed starch	Ruiloba et al, 2018.	Amylose/amylopectin: 1:2 Humidity:3.87% Good biodegradation	Casting
Tropical fruits	Limache Alonso and Limache López, 2018	Medium and good consistency They do not deform easily Good aesthetics Good shine Good and average duration	Thermocompressi on
tropical palm trees	Limache Alonso and Limache López, 2018	Very good consistency Very good rachis Little fragility Average size Good abundance	Thermocompressi on

As can be seen in Table 1, all authors carry out different reviews and the properties that each one reports are not necessarily the same as the other authors. However, a clear trend is seen with respect to tensile properties, since Half of them have reported at least one of these properties. It is also seen that flexural properties are important, since many of them also report results for these properties as well as for physical properties such as thickness and density. The same goes for chemical properties such as moisture content, solubility, and water vapor permeability. Some of the authors only present qualitative properties (marked with color), but most of them presented finished products in their research.

Stage 2: Annual production of the various raw materials

All of these plants mentioned are produced in Santander, and there are various sources of annual production that are presented. For cassava, various production data were found; for the year 2014, according to the Ministry of Agriculture in Santander, 110,538 tons were produced (Agronet, 2014), another data provided by the 2015 national agricultural survey, 517,489 tons of cassava were produced throughout Colombia. this tuber, but among the main producers the department of Santander is not named (DANE, 2016). According to a document published on the ICBF website, in the department of Santander for the year 2016, production was 118,011 tons (ICBF, 2016).

Regarding the cultivation of corn, the Ministry of Agriculture reported a production of 16,175 tons for Santander in 2014 (Agronet, 2014), and according to the ICBF document for the year 2016 the production was 16,493 tons (ICBF, 2106), showing that the production variation was not much. For pineapple, the Ministry of Agriculture reported a production of 218,269 tons in 2014 (Agronet, 2014) and for 2016, according to ICBF data, the production of this fruit was 378,975 tons (ICBF, 2016). while in 2018 for the department of Santander together with that of northern Santander, production was approximately 374,000 tons (Ministry of Agriculture, 2018), presenting an increase compared to 2014.

For potatoes, the 2015 national agricultural survey found a production of 99,249 tons (DANE, 2017), while in 2016 it was 53,477 tons (ICBF, 2016), which means a reduction in the cultivation of potatoes. this tuber. Regarding bananas, a production of 130,530 tons was reported in 2014 (Agronet, 2014) and in 2016 the report was 164,080 tons (ICBF, 2016), these figures revealing a slight increase in production.

Regarding rice, two data are presented for the year 2016, according to the 2016 rice census, 4,438 tons of rice were produced in Santander (Finagro, 2017), but according to the ICBF source (2016), the production was 7,571 tons. Oil palm had a production of 181,813 tons in

2014 (Agronet, 2014) and 200,180 tons in 2016 (ICBF, 2016), increasing production. Of cocoa, in 2014 production was 18,826 tons (Agronet, 2014), in 2016 production increased to 22,384 tons (ICBF, 2016) and in 2019 there was a slight increase to 23,000 tons (Ruiz, 2019).

With respect to wheat, a production of 110 tons was reported for 2016 and for coconut a production of 24 tons (ICBF, 2016), very small figures that would put the food security of the population at risk if these foods were used for manufacture other products.

Stage 3: Proposal for the manufacture of a biodegradable container for food storage based on the information analyzed

With the properties analyzed and the availability of the product, the final proposal was made. When using a product of natural origin for the manufacture of food packaging, if this is a product consumed by human beings, nutrition prevails over other uses that may be given to the product. For this reason and given the production figures of the different products in Santander, it is considered that the most optimal material to work with is the pineapple crown, since it is an agro-industrial waste that does not endanger people's food security.

But it is proposed to reinforce the material with one of the materials for which the Young's Modulus and water vapor permeability were reported, since according to Müller, González and Chiralt (2017), many of the products used to manufacture containers do not present all optimal properties, so a bilayer combination of two different materials with complementary properties can generate a material with optimal conditions.

In this sense, the final proposal consists of manufacturing a material with a pineapple crown according to the method of Kroefly, Robles and Vargas (2018), inserting an intermediate layer of the material made by Díaz (2015), which used corn starch and gelatin. bovine and obtained a Young's Modulus of 512 MPa and a water vapor permeability of 5 *10^-7g/Pa*s*my although it is not the best Young's modulus consulted, it is a good value. Polylactic acid, which had the best Young's moduli, is not taken into account; it is a long process to extract this material from starch, so the yields will not be as high. Nor is the combination of fique and cassava taken into account, which although it had a better Young's modulus than the material manufactured by Díaz, this study has made great progress and the material is already patented.

To prepare the material manufactured by Díaz (2015), the casting method is recommended, since with this method the water vapor permeability yielded lower values and once the two products were obtained, mold and press them with the help of heat.

Developing biodegradable containers for food storage, replacing conventional materials such as polystyrene and polypropylene, directly contributes to environmental sustainability in its relationship with the following sustainable development goals: *Goal 6: Clean water and sanitation, Goal 12: Responsible production and consumption , Goal 14: Underwater life and Goal 15: Life in terrestrial ecosystems.*

Each of these objectives has specific goals and being able to replace these conventional materials would allow a reduction in the generation of single-use waste, which ultimately ends up being disposed of in landfills or as waste that directly affects water resources such as streams, rivers, oceans or on land, which generates adverse impacts on human health and the environment. The goal is to protect and restore water-related ecosystems, including forests, mountains, wetlands, rivers, aquifers and lakes, as well as significantly reduce waste generation through prevention, reduction, recycling and reuse activities. , which would reduce the degradation of natural habitats, stop the loss of biological diversity and protect threatened species and prevent their extinction.

With the proposal to manufacture food containers from the use of waste, the development of an unrecognized business opportunity is generated, from which it contributes to the sustainable management of the environment and provides job opportunities that allow contributing to the value chain. of the natural resource used. For the business or commercial sector, this type of production is an opportunity to improve its social image, after projecting commitment to the environmental component and appropriating solutions that alleviate the problem of solid waste, in addition to adopting sustainable practices that would allow production and consumption. responsible.

In Santander there is a great variety of products of natural origin such as corn, cassava, potatoes, pineapple, mango, sapote, sugarcane and production of cattle and poultry, among others, with which food containers can be manufactured that replace those that currently exist since these are slow to degrade and highly polluting. All the products developed by the consulted authors are biodegradable and represent, in the medium and long term, a real solution to the problem generated by expanded polystyrene and other plastics used in food containment.

Many properties can be evaluated for a food container, but among the most important for its operation are mechanical properties such as tensile strength, Young's modulus, elongation percentage, bending resistance, load maximum support and hardness. Also chemical properties such as water vapor permeability, solubility, glass transition temperature and degradation start temperature, since these are what indicate how functional the manufactured material can become. Regarding the processes for manufacturing containers, they are diverse, but the most notable are the casting or *casting method,* molding with pressure and heat or thermoforming, and the extrusion process.

Pineapple, as one of Santander's flagship products, represents an alternative for the production of biodegradable food containers, since its crown, which is normally wasted. This can be combined with other materials such as corn starch and bovine gelatin to obtain a material with better properties and that in the future will be a solution to the problem of single-use food packaging. And it is the most viable option taking into account that the other raw materials do not endanger people's food security.

REFERENCES

Agronet. (2014). Santander. Main Crops by Area Planted in 2014. Retrieved from: http://www.agronet.gov.co/Documents/Santander.pdf

Alarcón, H. and Arroyo, E. (2016). Soc Quim Peru Magazine. Evaluation of the chemical and mechanical properties of biopolymers from modified potato starch, 82(3). [315-323]. Retrieved from: http://www.scielo.org.pe/pdf/rsqp/v82n3/a07v82n3.pdf

Arévalo, K., Alemán, M., Rojas, M. and Morales, L. (2010). Latin American Journal of Environmental Biotechnology Algal. Biodegradable films from citrus waste: proposal for active packaging 1(2). [124 – 134]. Retrieved from: https://www.virtualpro.co/biblioteca/peliculas-biodegradables-a-partir-de-residuos-de-citricos-propuesta-de-empaques-activos

Atencio, V., Gallego, L., Torreblanca, D. and Zuleta, A. (2018). Obtaining a composite material of natural origin. (Graduate work, Universidad Pontificia Bolivariana). Retrieved from: https://repository.upb.edu.co/handle/20.500.11912/4265

Avalos, A. and Torres, I. (2018). Business model for the production and marketing of biodegradable packaging based on rice husks. (Graduate work, University of Piura). Retrieved from https://pirhua.udep.edu.pe/handle/11042/3459

Ayala, A. and Villada, H. (2015). Biotechnology in the agricultural and agroindustrial sector. effect of cassava flour gelatinization on the mechanical properties of bioplastics, 13(1). [38-44]. Retrieved from: https://www.researchgate.net/publication/283635858_Efecto_de_la_gelatinizacion_de_la_harina_de_yuca_sobre_las_propiedades_mecanicas_de_bioplasticos/link/58a5cb70aca27206d98f5dae/download

Batuani, R. (2015). Study of obtaining biodegradable plastics from potato starch by adding plasticizing agents. (Graduate thesis, Universidad Mayor de San Andrés). Recovered from: https://repositorio.umsa.bo/handle/123456789/9295

Bejarano, N. (2018). Study of the mechanical properties of a biopolymer based on the starch content of banana peel. (Graduate thesis, National University of San Agustín de Arequipa). Retrieved from: http://bibliotecas.unsa.edu.pe/bitstream/handle/UNSA/7578/MTbemanl.pdf?sequence=3&isAllowed=y

Canaza, R. (2012). evaluation of biodegradable films obtained from oxidized starch of banana (Musa paradisiaca) variety inguri. (Graduate thesis, National University of the Altiplano). Recovered from: http://repositorio.unap.edu.pe/handle/UNAP/3382

Cárdenas, D. and Medina, J. (2011). Compound based on cassava starch and coffee cisco, for the production of disposable food containers. (Graduate thesis, Universidad de los Andes). Recovered from: https://repositorio.uniandes.edu.co/handle/1992/14836

Carvajal, S. (2019). Obtaining biodegradable packaging from collagen and starch. (Graduate work, University of the Americas). Retrieved from: http://dspace.udla.edu.ec/bitstream/33000/11588/1/UDLA-EC-TIAM-2019-33.pdf

Castellón, H. (sf). Oxo-biodegradable plastics vs. Biodegradable plastics: what is the way? Recovered from http://files.udescesos.webnode.es/200000042-df18fe0252/1_HELLO_CASTELLON.pdf

Chacón, T (2019). Study of the properties of teak and sapote leaves for the production of disposable food packaging. (Master's thesis, La Molina National Agrarian University). Recovered from: http://repositorio.lamolina.edu.pe/handle/UNALM/4189

Chariguaman, J. (2015). Characterization of starch bioplastic prepared by the casting method reinforced with passion fruit albedo (passionflower edulis spp.). (Graduate thesis, Pan American Agricultural School, Zamorano). Retrieved from: https://bdigital.zamorano.edu/bitstream/11036/4560/1/AGI-2015-014.pdf

Charro, M. (2015). Obtaining biodegradable plastic from potato starch. (Graduate thesis, Central University of Ecuador). Retrieved from: http://www.dspace.uce.edu.ec/handle/25000/3788

Corbella, C., Hernández, M. and Laz, M. (2018). ECOPLATES. Disposable plates made from plant waste. (Graduate thesis, University of La Laguna). Recovered from: https://riull.ull.es/xmlui/handle/915/10258

Cruz, J., Cueva, F., García, M., Gudiel, A. and Sigüenza, Y. (2018). Design of a production plant to obtain biodegradable dishes based on coconut tow in the province of Piura. (Graduate work. University of Piura). Retrieved from: https://pirhua.udep.edu.pe/bitstream/handle/11042/3838/PYT_Informe_Final_Proyecto_PLATOSBIODEGRADABLES.pdf?sequence=1&isAllowed=y

DANE. (2016). The cultivation of cassava (Manihot sculenta Crantz). Retrieved from: https://www.dane.gov.co/files/investigaciones/agropecuario/sipsa/Bol_Insumos_abr_2016.pdf

DANE. (2017). The cultivation of potato (Solanum tuberosum L.) and a case study of the production costs of Pastusa Suprema potato. Retrieved from: https://www.dane.gov.co/files/investigaciones/agropecuario/sipsa/Bol_Insumos_ene_2017.pdf

Díaz, R. (2015). Biodegradable antimicrobial films based on starch and gelatin. (Master's thesis, Polytechnic University of Valencia). Recovered from: https://riunet.upv.es/bitstream/handle/10251/56543/D%c3%8dAZ%20-%20FILMS%20BIODEGRADABLES%20ANTIMICROBIANOS%20A%20BASE%20DE%20ALMID%c3%93N%20Y% 20GELATIN.pdf?sequence=2&isAllowed=y

Echeverría, I. (2012). Biodegradable materials based on soy proteins and montmorillonites. (Doctoral thesis, National University of La Plata). Retrieved from: http://sedici.unlp.edu.ar/handle/10915/18534

Elnoticom (April 17, 2018). Goodbye to plastic plates, in Colombia they make them with banana leaves. Recovered from: https://elnoti.com/adios-los-platos-plasticos-colombia-los-fabrican-hojas-platano/

Espina, M., Cruz-Tirado, JP and Siche, R. (2016). Scientia Agropecuaria. Mechanical properties of trays made with starch from native plant species and fibers from agroindustrial waste 7(2). [133-143]. Retrieved from: http://www.scielo.org.pe/scielo.php?pid=S2077-99172016000200006&script=sci_abstract

Finagro. (2017). Rice – Irrigation. Agroeconomic Reference Framework. Retrieved from: https://www.finagro.com.co/sites/default/files/node/basic-page/files/arroz_riego.pdf

García, AV (2015). Obtaining a biodegradable polymer from corn starch. (Graduate work, ITCA–FEPADE Specialized School of Engineering). Retrieved from https://www.itca.edu.sv/wp-content/themes/elaniin-itca/docs/2015-Obtencion-de-un-polimero-biodegradable.pdf

García, Y., Zamudio, P., Bello, L., Romero, C. and Solorza, J. (2011). Ibero-American Journal of Polymers. Oxidation of native banana starch for its potential use in the manufacture of biodegradable packaging materials: physical, chemical, thermal and morphological characterization, 12(3). [125 -135]. Retrieved from: https://www.virtualpro.co/biblioteca/oxidacion-del-almidon-nativo-de-platano-para-su-uso-potential-en-la-fabricacion-de-materiales-de-empaque-biodegradables -physical-chemical-thermal-and-morphological-characterization

García, L., García, A., Olaya, P., Rosas, G. and Vignolo, D. (2019). Design of the production process of biodegradable trays from cornstarch. (Graduate thesis, University of Piura). Retrieved from: https://pirhua.udep.edu.pe/bitstream/handle/11042/4276/f15f56df2c6ce4fa6b1beb82a733acaee5e1247a3a2fc682d78384751f8c7955.pdf?sequence=1&isAllowed=y

Giraldo, L., González, G., Ochoa, M. and Tello, E. (2018). Development of sustainable containers using oranges as raw material. (Graduate thesis, Universidad Pontificia Bolivariana). Retrieved from: https://repository.upb.edu.co/handle/20.500.11912/4215

González, J. and Tovar, M. (2018). Development of a biodegradable polymer from ataco seed starch, Amaranthus quitensis L. (Doctoral thesis, Universidad Nacional Mayor de San Marcos). Retrieved from: http://cybertesis.unmsm.edu.pe/handle/cybertesis/9535

Santander Government. (March 6, 2020). The Always Santander Government decreed a ban on the use of plastic and Styrofoam. Retrieved from: http://www.santander.gov.co/index.php/actualidad/item/4652-gobierno-de- siempre-santander-decreto-la-prohibicion-del-uso-de-plastico-e-icopor

Holguín, J. (2019). Obtaining a bioplastic from potato starch. (Graduate thesis, Universidad de América Foundation). Retrieved from: http://repository.uamerica.edu.co/bitstream/20.500.11839/7388/1/6132181-2019-1-IQ.pdf

ICBF. (2016). Department of Santander. crops-production in tons/year. Retrieved from: https://www.icbf.gov.co/sites/default/files/oferta_agricola_-santander_2016.pdf

Iturbe, A. (2016). Disposable banana leaf container. (Master's thesis, Universidad Panamericana). Retrieved from: https://tesis.gdl.up.mx/100796.pdf

Kroefly, A., Robles, B. and Vargas, E. (2018). Expolbero Autumn. Prototype of a biodegradable disposable plate made from the crown of the pineapple (Ananas comosus). [1-3]. Recovered from: https://repositorio.iberopuebla.mx/handle/20.500.11777/4113

Limache Alonzo, A., & Limache López, AM (2018). Impact of tropical palm trees in obtaining ecological dishes. Recovered from: http://repositorio.unu.edu.pe/handle/UNU/4155

Limache Alonzo, A., & Limache López, AM (2018). Ecological dishes made from the leaves of 10 species of tropical fruit trees. Recovered from: http://repositorio.unu.edu.pe/handle/UNU/4156

López, BI and Fajardo, JL (2019). Design of biodegradable containers for food transportation. (Graduate work, University of Azuay). Retrieved from http://dspace.uazuay.edu.ec/handle/datos/9182

López, S. and Largo, J. (2018). Verde Green – applied environmental engineering for the production and marketing of biodegradable disposables. (Graduate work, Autonomous University of the West). Retrieved from: http://red.uao.edu.co:8080/bitstream/10614/10614/5/T08280.pdf

Luna, G., Villada, H. and Velasco, R. (2009). Dyne. Thermoplastic cassava starch reinforced with fique fiber: preliminaries (76). [145-151]. Retrieved from: http://www.scielo.org.co/scielo.php?script=sci_abstract&pid=S0012-73532009000300015&lng=e&nrm=iso

Mayorga, J., Rodríguez, L. and Martínez, J. (2018). Biodegradable banana leaf packaging. case study healthy combo program - Industrial University of Santander. (Graduate thesis, Industrial University of Santander). Retrieved from: http://noesis.uis.edu.co/handle/123456789/28311

Meza, P. (2016). Preparation of bioplastics from residual starch obtained from potato peelers and determination of its biodegradability at laboratory level. (Graduate work, La Molina National Agrarian University). Recovered from: http://repositorio.lamolina.edu.pe/bitstream/handle/UNALM/2016/Q60-M49-T.pdf?sequence=1&isAllowed=y

Department of agriculture. (2018). Pineapple production would reach more than 950 thousand tons in 2018, estimates MinAgricultura. Recovered from: https://www.minagricultura.gov.co/noticias/Paginas/Producci%C3%B3n-de-pi%C3%B1a-llegar%C3%ADa-am%C3%A1s-950-mil-toneladas -in-2018,-calcula-MinAgriculture-.aspx

Morales, C. (December 1, 2019). Single-use plastics will be banned in seven Caribbean countries. The FM. Retrieved from: https://www.lafm.com.co/medio-ambiente/plasticos-de-un-solo-uso-seran-prohibidos-en-siete-paises-del-caribe

Muñoz, J. (2014). Evaluation, characterization and optimization of a bioplastic from the combination of corn starch, cassava and glycerol in its physical and barrier properties. (Graduate thesis, Pan American Agricultural School, Zamorano). Retrieved from: https://bdigital.zamorano.edu/bitstream/11036/3366/1/AGI-2014-T033.pdf

Narváez, MA (2016). Optimization of the mechanical properties of bioplastics synthesized from starch. (Graduate thesis, Universidad San Francisco de Quito). Recovered from: http://repositorio.usfq.edu.ec/handle/23000/6299

Navia, D. (2011). Development of a material for food packaging from cassava flour and fique fiber. (Master's thesis, Universidad del Valle). Retrieved from: http://bibliotecadigital.univalle.edu.co/bitstream/10893/8845/1/TESIS%20MAESTR%C3%8DA%20Diana%20Navia.pdf

Navia, D., Ayala, A. and Villada, H. (2015). Technological information. Cassava Flour Biocomposites obtained by Thermocompression. Effect of Process Conditions 26(5). [55-62]. Recovered from : https://scielo.conicyt.cl/pdf/infotec/v26n5/art08.pdf

Ortega, J. (2016). Study of the physicochemical and functional properties of banana flour (Musa acuminata AAA) for rejection in the development of biodegradable films. (Graduate thesis, Technical University of Ambato). Retrieved from: https://repositorio.uta.edu.ec/bitstream/123456789/22874/1/AL599.pdf

Pardo, O. and Velasco, R. (2011). Ion Magazine. Physicochemical and mechanical properties of films obtained from native and oxidized arracacha starch, special edition (2012). [23 – 29]. Recovered from: https://www.virtualpro.co/biblioteca/propiedades-fisicoquimicas-y-mecanicas-de-peliculas-obtenidas-a-partir-de-almidon-nativo-y-oxidado-de-arracacha

Pizá, H., Rolando, S., Ramírez, C., Villanueva, S. and Zapata, A. (2017). Experimental analysis of the production of bioplastic from banana peel for the design of an alternate production line for the chifleras of Piura, Peru. (Graduate thesis, University of Piura). Retrieved from: https://pirhua.udep.edu.pe/bitstream/handle/11042/3224/PYT_Informe_Final_Proyecto_Bioplastico.pdf?sequence=1&isAllowed=y

Pérez, I. and Ramos, L. (2019). Use of waste from the paradisiacal muse (banana) to obtain biodegradable packaging. (Graduate work, José Faustino Sánchez Carrión National University). Recovered from: http://repositorio.unjfsc.edu.pe/handle/UNJFSC/3503

Pretell, M. (2018). Manufacture of edible tableware from biopolymers of cassava (Manihot sculenta) and aloe vera (Aloe vera). (Graduate thesis, Professional School of Environmental Engineering). Recovered from: http://repositorio.ucv.edu.pe/handle/UCV/21091

Puello, B. and Zabaleta, L. (2014). Obtaining a biodegradable film from corn cobs to be used as food packaging, on a laboratory scale, at the University of San Buenaventura Cartagena. (Graduate thesis, University of San Buenaventura). Retrieved from: http://bibliotecadigital.usbcali.edu.co/bitstream/10819/2620/1/Obtenci%C3%B3n%20de%20una%20pel%C3%ADcula%20biodegradable_Brenda%20Puello_USBCTG_2014.pdf

Quintanilla, M. (2010). Industrialization of coconut tow fiber. (Graduate thesis, University of El Salvador). Retrieved from: http://ri.ues.edu.sv/431/1/10136579.pdf

Ranaldi, A. Roldan, M. Urdinola, D. (sf). Polymeric food. (Graduate work, Universidad Pontificia Bolivariana). Retrieved from: https://repository.upb.edu.co/bitstream/handle/20.500.11912/3837/Alimento%20Polim%C3%A9rico.pdf?sequence=1

Vida Editorial (March 27, 2019) How long does it take for plastics to decompose? TIME. Retrieved from: https://www.eltiempo.com/historias-el-tiempo/cuanto-tiempo-tardan-los-plasticos-en-descomponerse-342568

Ríos, J. (March 1, 2020). La Belleza, Santander, prohibits the use of plastic in public contracts. Liberal Vanguard. Recovered from: https://www.vanguardia.com/santander/guanenta/la-belleza-santander-prohibe-el-uso-de-plastico-en-contratos-publicos-AN2078406

Ruiloba, I., Li, M., Quintero, R., & Correa, J. (2018). Magazine of Scientific Initiation. Preparation of bioplastic from mango seed starch, 4, 28-32. Retrieved from: https://pdfs.semanticscholar.org/801b/86fd684502d3d69e67f1b81f3b44bcabc284.pdf

Ruiz, L. (November 1, 2019). Cacao de Santander continues to stand out in production and flavor. Liberal Vanguard. Retrieved from: https://www.vanguardia.com/economia/local/cacao-de-santander-sigue-destacandose-en-produccion-y-sabor-ym1619504

Salgado, V, C. and Granja, W. (2016). Analysis of the promotion and possible application of 100% biodegradable packaging for natural snacks. (Graduate work, University of the Americas). Recovered from: http://dspace.udla.edu.ec/handle/33000/5489

Tejada, C., Tejeda, LP, Tejeda, LM and Villabona, A. (2007). Teknos. Polylactic acid: A biodegradable plastic made from starch. Recovered from: https://dialnet.unirioja.es/servlet/articulo?codigo=6382678

Velasco, R., Enríquez, M., Torres, A., Palacios, L. and Ruales, J. (2012). Biotechnology in the Agricultural and Agroindustrial Sector. Morphological characterization of biodegradable films from modified cassava starch, antimicrobial agent and plasticizer, 10(2). [152 - 159]. Recovered from: https://www.virtualpro.co/biblioteca/caracterizacion-morfologica-de-peliculas-biodegradables-a-partir-de-almidon-modificado-de-yuca-agente-antimicrobiano-y-plastificante

Zarate, L. (2011). Biodegradable polymeric materials prepared by thermomechanical processing from gluten/plasticizer mixtures. (Graduate thesis, University of Seville). Recovered from: https://dialnet.unirioja.es/servlet/tesis?codigo=24079

CAPITULO 3.
ANALISIS DE VULNERABILIDAD AL CAMBIO CLIMÁTICO DESDE LAS COMUNAS DE BUCARAMANGA, COMO HERRAMIENTA PARA LA SOSTENIBILIDAD DE LA CIUDAD

CHAPTER 3.

ANALYSIS OF VULNERABILITY TO CLIMATE CHANGE FROM THE COMMUNES OF BUCARAMANGA, AS A TOOL FOR THE SUSTAINABILITY OF THE CITY

Carlos Alberto Amaya Corredor

Cadastral and Geodesta Engineer, Specialist in Planning for Environmental Education, Master in Environmental Management and Audits, M.Sc. Sustainable Development and Environment

Research professor of the GIECSA Research Group, assigned to the Environmental Engineering program of the Technological Units of Santander

camaya@correo.uts.edu.co

SUMMARY

Climate change is the environmental situation with the greatest global impact. Since 1994, the United Nations Framework Convention on Climate Change has refocused efforts to look not only at the consequences, but also at combating the causes of climate change and transmitting to communities initiatives that stop negative actions for the planet. The world vision today focuses on training the population in three key aspects: adaptation to new ecosystem dynamics and new ways of life; mitigation of the damages and impacts that are generated and citizen education to understand environmental changes. In this context, this project has focused on identifying the city factors that affect the mitigation, adaptation and knowledge of climate change, so that, recognizing reality, decisions can be made for environmental protection and conservation and strengthening the urban living environment. For this, work was done to recognize the city's vulnerability conditions, with a unit of analysis in its division by urban communes. A qualitative criterion of level of vulnerability was established, in which to identify each commune when crossing: international criteria of vulnerability and conditions of sensitivity, exposure and information, this relationship gender classification levels of each commune. It was established that Bucaramanga has a medium level of vulnerability to climate change, given that in aspects such as the average temperature of the city is constant and it has good green areas, public spaces, an image of a friendly city and willingness of the inhabitants to help improve the environment, but presents difficulties in mobility and solid waste management, which significantly deteriorates the urban ecosystem. This analysis coincides with the IDEAM approach on climate change and risk scenarios in Colombia and in the perspective of the departmental climate change plan, Santander 2030. The medium level of Vulnerability to Climate Change calls for strengthening citizen training actions against to environmental issues, as well as greater government commitment to strengthening the city's natural coverage, improving solid waste management and sustainably regulating the city's vehicular traffic. Finally, a call is made to the business sector to recognize its role in the management of actions to mitigate climate change, after appropriating the issue, generating particular adaptation and training actions and supporting public authorities and the general population in executing actions that counteract climate change, all hoping that, with a better environmental image, greater social recognition that will be reflected in better income from the marketing of their products.

INTRODUCTION

The United Nations Framework Convention on Climate Change (UNFCCC) (CMNUCC, 1994), in its article 1, defines climate change as "climate change attributed directly or indirectly to human activity that alters the composition of the global atmosphere and that adds to "natural climate variability observed over comparable time periods"

In the chapter for Central and South America, of the fifth report on climate change, the Intergovernmental Panel on Climate Change (IPCC) (Magrin, 2014)documents the importance of social practices that contribute to the management of adaptation to climate change, after evidencing and seeking ways to manage the different vulnerabilities that a society has. The vulnerability of urban spaces depends on their ability to integrate urban conditions and ecological conditions in the space, so that the effects of climate change can be attenuated and managed.

In Europe, most cities have very advanced adaptation plans (AEMA, 2015), developing with special attention elements around storms, floods and heat waves, as the main agents affecting European cities due to the phenomenon of climate change. The processes of Rotterdam, Ghent and Bologna stand out, in which they have worked on technical processes to demonstrate climate change in cities and social processes to educate the population in the identification of the phenomenon and self-management to mitigate or adapt to changing scenarios.

Furthermore, Europe has a good sample of cities oriented by climate change adaptation and mitigation plans. Växjö (Sweden) (INI, 2016), shows the consolidation of the city, with a strong component of green coverage and the benefits of this in the face of climate change; Copenhagen (Denmark) (Sengupta, 2019)is shown as a city in which government guidelines are aimed at establishing control and mitigation measures for climate change, identifying the endogenous conditions of the city; On the other hand, Amsterdam (Holland) (Heras, 2015), makes a multidimensional commitment for its inhabitants to reconvert their lifestyles, incorporating considerations of adaptation and mitigation to climate change; In Spain, the cities San Sebastián (GEA21, 2017), Valencia (Valencia, 2015), Murcia (Fonfria, 2017), Barcelona (Barcelano, 2020), are committed to managing climate change conditions, following the prospective vision of the cities, preserving their historical heritage, strengthening their possibilities for development and construction of infrastructure, incorporating certain types of coverage. greens in cities and promoting citizen participation as a central element of the city vision.

All of Europe is affected by climate change (Breil et al., 2018), strong evidence of which can be periodic large "Heat Waves" which affect all urban activities and dynamics, forcing social intervention measures to be taken to help inhabitants manage and overcome the problems. these unexpected weather periods.

At the level of the American continent, Curitiba (Brazil) (UNICC, 2017)develops tools for adaptation and mitigation to climate change, among which its commitment to sustainable mobility and the recomposition and strengthening of green coverage and green areas for the enjoyment of its inhabitants stands out; Brasilia (Brazil), is one of the youngest cities in Latin America, it has been designed and built with urban planning and territorial ordering strategies that contemplate the balance between nature, forest cover, future developments and city infrastructure; Quito (Ecuador) (ELLA, 2013)stands out as a Latin American city committed to developing response strategies to climate change, which have focused on combining risk reduction with the development of institutional capacity and greater citizen participation; for the city of Santiago de Chile (Chile) (Gobierno de Chile, 2017)from the National Plan for Adaptation to climate change that is contemplated for the city by the country's government strategies, highlighting the importance of the consolidation of natural spaces and green areas in the city and the inclusion of the inhabitant as the protagonist of adaptation and mitigation to climate change; The analysis made for Buenos Aires (Argentina) (Herrero y otros, 2018), focuses on the recognition, attention and solution of "socio-territorial" conflicts, which from the vision of the inhabitants achieve the reconstruction of the territories and consequently the generation of strategies and actions for adaptation and climate change mitigation.

In the National context, Colombia has built the National Plan for Adaptation to Climate Change that guarantees the bases adapted to the territory, for mitigation and adaptation to possible events and the gradual transformation that the country faces. Santander was part of the first six departments to formulate its Adaptation and Mitigation plan, the city of Bucaramanga has not yet taken this initiative.

By having a Country Plan, cities have taken initiatives to manage their climate change incidents. Bogotá, as the capital of the country, has incorporated adaptation and mitigation into its policies (Bogota, 2015); Medellín is a benchmark for sustainable urban management not only at the country level, but in America, its management of public spaces and green corridors has provided it with strengths in response to climate change (Medellin, 2015); The city of Cali articulated the municipal development plan with an adaptation and mitigation plan, which is a strength for the quality of urban life of its inhabitants (Cali, 2020); Adaptation and mitigation measures are taken from Barranquilla, articulated with the departmental climate change plan, Atlántico 2040, which focuses on actions for the entire departmental territory, seeking comprehensiveness of the actions and positively impacting its entire population (MinAmbiente & CAEM, 2016); Intermediate cities like Ibagué, within their municipal development plan, incorporate actions for the adaptation and mitigation of climate change, in search of the well-being of their population (Alcaldía de Ibague, 2016). Likewise, Bucaramanga has been articulated within the departmental climate change management plan of the department and from there guide strategies and actions to achieve management of the territory in favor of the population.

Although the country has led regional efforts to adapt to climate change, there are still strong gaps in the analysis of urban territories. Only from the regional strategy of emerging and sustainable cities, promoted by the Inter-American Development Bank (IDB), issues such as climate change have taken on special relevance in city planning. The evaluation of the program for Bucaramanga, until 2015 and its projection to the 2019 period, gives special importance to the climate change component in city planning in its construction under sustainability criteria (BID, 2016), which was enhanced and highlighted as a component. importance of the city's sustainability, in the 2020-2023 municipal development plan, making a special description of the conditions that exist in the city of Bucaramanga.

It is therefore of total importance for the management of climate change, to identify the urban processes that are related and from which the risk factors that the urban configuration generates, the climatic conditions that threaten the consolidation of urban life can be cataloged and for this it is It is essential to identify the city structure to establish its vulnerability condition, thereby knowing the strong aspects or those with the best response and the weak aspects or those most affected by the dynamics of climate change. The analysis of this vulnerability fundamentally involves recognizing the physical and social aspects of the territory, understanding that the economic components are marked by the direction of public investment that the municipal administration prospects in each four-year term of Mayor.

In accordance with the consolidated data presented by ECLAC (Margulis, 2016), in cities variables such as heat island, increase in vehicle fleet, solid waste management, decrease in green areas, training and citizen education can be identified, from which conditions that explain the exposure are recognized. , sensitivity and information and capacity to adapt to climate change scenarios and with which the condition of a city in the face of climate change can be catalogued. This analysis has been applied in the city of Bucaramanga, taking as study environments, subdivision by communes, zones of social subdivision, to administratively focus the territory, from which the community-territory relationship facilitates the recognition of its own conditions and the level d citizen commitment to contribute to the improvement of the entire city, so that planning, training and territorial management decisions can be made that contain climate change in favor of the quality of life of the inhabitants.

IDENTIFIED PROBLEM

ECLAC (Margulis, 2016)has managed to manage important information on the vulnerability of Latin American cities to climate change, thereby presenting a set of variables from which the situation of the city in the face of climate change can be studied and search for mitigation alternatives. , mitigation and adaptation. It presents important considerations to advance adaptation and mitigation to climate change, emphasizing mitigation as a global commitment and adaptation as a strategy applicable at the local level to rebuild the lifestyles of regions and cities.

In this study, ECLAC, in accordance with the IPCC, analyzes how global risks of climate change are concentrated in urban areas. Heat stress, extreme rainfall, coastal flooding, landslides, air pollution, drought and water scarcity pose risks in urban areas to people, assets, economies and ecosystems.

According to a study carried out by the University of Manizales, through the Environment and Sustainable Development Research Center CIMAD, on the situation of vulnerability to climate change for the Santanderes region; The results show that the community is aware of climate variation, 63.4% of those surveyed say they know the policies aimed at reducing vulnerability to climate change in their region.

These policies are based on adaptation activities to the variability of climate change, risk management plan, campaigns aimed at environmental education, national development plan, departmental development plan and municipal development plan. As a follow-up to this research, the comprehensive territorial climate change management plan for the department of Santander predicts that during the period 2011 - 2040 there will be an increase in average temperature of up to 0.9 °C and an increase in precipitation of up to 0.54%, compared to the average records of the reference period 1976 – 2005.

For this reason, the Institute of Hydrology, Meteorology and Environmental Studies IDEAM, within the framework of the third national communication on climate change (TCN) (IDEAM I. d., 2016), carried out a vulnerability analysis for the year 2040 assuming that the adaptation conditions are the same as the current ones based in 84 indicators, where it can be seen that the threat is configured with a higher percentage of contribution by the components of: food security (66.81%), human habitat (10.55%) and water resources (10.24%) and the The highest threat values are given in the biodiversity component (0.87%).

As a complement to the consequences of climate change in Santander, the analysis of vulnerability and risk due to climate change in Colombia (IDEAM I. d., 2016)classifies Bucaramanga as one of the municipalities with high risk due to this phenomenon that has come to alert the world. So it is important to analyze the conditions with which the city's urban environment can face the growing conditions of global and regional climate change. This phenomenon affects the Bumanguesa population every day, due to the high temperatures, health is at risk, on the other hand, the ecosystems are affected by the risk of loss of species, the fertile areas of the city would be exposed to desertification, the which leads to losses in Santander agriculture.

The identification of the effects of climate change must be done at the level closest to the daily life of the city, to do so, reviewing the conditions of each commune of the city of Bucaramanga is a path that helps to detail the city, with the vision of its inhabitants, recognizing its physical urban elements and identifying the green spaces that make up the city, so that climate change is identified and understood in nearby scenarios, which encourages decision-making, the execution of actions and the resilient transformation of the city, knowing its vulnerabilities to anticipate the risks and threats that climate change throws on the global context.

CHAPTER DEVELOPMENT

The construction of a concept of approximation to the conditions of vulnerability of the city to climate change was formulated from the general to the particular, to propose better life scenarios. To do this, taking the city communes as the territorial study space, the general context of threat and risk criteria in the face of climate change in urban environments was recognized, with which to establish the conditions of vulnerability, which may occur in each commune of the city and propose strategies and actions that reduce this vulnerability and promote better mitigation and adaptation capacities to climate change.

METHODOLOGY

The research project focused on correlation (Hernandez y otros, 2014), under which it seeks to know the relationship that exists between the conditions and characteristics of urban life and how they behave, contributing or avoiding, consequences or alterations associated with climate change. With the information obtained, proposals for adaptation and mitigation strategies to climate change were constructed, which, if applied in the area under study, would help transform the lifestyles of the inhabitants and consequently transform the structure of the urban sector. , strengthening it against the effects that the global process of climate change may cause.

To build possible adaptation and mitigation scenarios, information has been collected from reference cities at the global, Latin American and country levels, from which to recognize how to identify the effects that climate change generates in urban environments and in the same way document the adaptation and mitigation strategies that have been applied for urban sustainability. This information will allow listing possibilities to be applied, which must be refined to the extent that the conditions that have structured the city of Bucaramanga are also documented. The relationship between cities and climate change has extensive documentation to identify these conditions of climate change incidents. From ECLAC, urban responses to climate change (Sánchez, 2013); the description and analysis of vulnerability and adaptation of cities in Latin America (Margulis, 2016)and the analysis of the relationship between climate change and urban sustainability (Galindo y otros, 2015), as well as from the European Environment Agency (EEA), ten cases of urban studies of how Europe adapts to climate change (Agencia Europea de Medio Ambiente, AEMA, 2018), they allow us to recognize elements, variables and situations that in cities are related to changes and effects derived from climate change.

Information was taken from secondary sources of information, fundamentally the territorial public authorities: the Municipal Mayor's Office of Bucaramanga, the environmental authority: Autonomous Corporation for the Defense of the Bucaramanga Plateau CDMB, the territorial link of the metropolitan region: Metropolitan Area of Bucaramanga AMB and the private entity: Bucaramanga Public Improvements Society; to socioeconomically characterize the population and identify city icons and public infrastructure, so that with the criteria of ECLAC and EEA, identify the urban life processes that may be contributing to the alterations and effects generated by climate change in the area . Once the conditions of the study area are recognized, they are compared with information from the reference cities and experiences, to establish what conditions of an urban environment are occurring in the city commune by commune. This allows us to focus the analysis on the endogenous characteristics of the territory to enhance their existence as elements of adaptation and mitigation to climate change.

Recognizing the incidence of climate change in urban environments, it is necessary to establish the conditions of these urban environments to respond, oppose or overcome the incidences of climate change, this can be understood as the conditions of vulnerability of the environment, which is related to the exposure to a threat, the sensitivity to it and the ability to adapt of those who are exposed. Chavarro (2013) exposes the guidelines with which the UN and the Colombian Ministry of Environment guide the description of vulnerability to climate change (Chavarro et al., 2013). Once the vulnerability conditions were appropriate, four possible levels of vulnerability were established: high, medium, low and not vulnerable; with which to describe the different effects of the city in the face of climate change.

With what was identified from ECLAC and EEA, four physical variables possible to analyze in the urban environment of the city of Bucaramanga were identified (the compilation indicated between 12 and 18 variables when analyzing cities and the environment, but due to the characteristics of the city only the incidence of four of them was possible) : Islands or heat waves, increase in the number of vehicles, solid waste management, decrease in green areas.

With the adaptation methodology developed by Amaya (2019), based on the green cities index proposal of the multinational SIEMENS (2010), a matrix analysis was carried out between environmental vulnerability conditions and the selected physical variables, to identify by communes. , what levels of vulnerability are generated and with this, unify city vulnerability conditions, from which to focus on prospective decision-making that strengthens the city's resilience to climate change. (Amaya y otros, 2019),(EIU, 2010)

Finally, intervention strategies are proposed, which are proposed from the urban adaptation and mitigation experiences, so that they enhance the characteristics of the study area. These strategies will be based on the experiences of cities (Agencia Europea de Medio Ambiente, AEMA, 2018), (Galindo y otros, 2015) (Breil et al., 2018)which show adaptation and mitigation results and that are possible to be applied proactively in the study area, the intervention in the urban infrastructure of the city will be sought to be minimal, to reduce costs associated with these processes and focus actions on the inhabitants and the sustainable remake of their customs and life dynamics.

To cover the entire city, from the Research Group on Ecosystems and Environmental Services GIECSA, from the Environmental Engineering program of the Technological Units of Santander, we worked with students in their last semester, with conditions met to advance their degree project, to analyze the city of Bucaramanga, in the face of climate change, according to its division by urban-administrative communes, for which 15 work groups were determined that would allow detailed tours and targeting of each commune of the city. The degree reports from each of these 15 work groups made it possible to consolidate the information on the city later, from the research group, to unify the criteria for analyzing conditions, urban physical variables and levels of vulnerability.

RESULTS AND DISCUSSION

From the base documents, the elements were established from which to describe urban conditions and their relationship with climate change.

Taking as an initial reference the IPCC approach and the ECLAC guidelines, in the urban area, collective facilities, public spaces, housing and habitat, mobility, roads and transportation, cultural heritage, home public services, management and disposal of solid waste and increase of sea level, are characteristics that allow identifying and monitoring climate change.

The inhabitant is the one who determines the organization of the territory, which is why the social composition of urban spaces constitutes aspects of the city that show climate change as a condition of development, and, furthermore, as an element present in the analyzes of sustainable development that must be guide cities to build quality of life for their inhabitants. Here aspects such as the adaptation and recovery capacity of communities, the importance of gender integration and the aspects associated with socio-environmental processes present in the incidence of climate change in urban environments are highlighted. (Novillo y otros, 2018).

Special importance is placed on the need to work in cities on aspects of citizen education, knowledge of the issue of climate change in the daily lives of inhabitants, the integration of the community in decision-making and the mechanisms that strengthen knowledge of climate change as a condition of city conditions. In the particular case of the city in Colombia, they emphasize the importance of evidencing climate change in all aspects of city construction, to demonstrate it not only associated with economic conditions but with all the conditions that influence the quality of urban life.

Urban spaces on the American continent have been oriented along these lines of interpretation. The cities of the continent have similar aspects, the physical configuration of the space, in which their distribution is recognized from a historical, religious and administrative center, from where the available space was populated. (Galindo y otros, 2015). In this configuration, the cities, variables such as housing infrastructure and urban space, show similar age and decrease in greenery to the extent that the city condenses in its original center, to have new spaces on its periphery with green areas limited to public mobility spaces. The coverage of public services reaches more than 95% of the cities and solid waste management is a conflictive issue in all the documents taken as a source, although experience such as Brazil (Niemeyer & Costa, 2018)shows significant improvements and efforts such as those in Mexico City (Delgado y otros, 2015)have achieved show progress, the trend is general in America.

For its part, AEMA has consolidated a clear list of the elements that must be taken from an urban space, for the management of climate change (Agencia Europea de Medio Ambiente, AEMA, 2018), classifying sectors into: agriculture, biodiversity, buildings, coastal zones, disaster risk reduction, energy, forestry , finance, health, urban transportation and water management; and with these sectors, impacts have been identified that affect, transform and deteriorate urban life: extreme temperatures or heat waves, floods, rising sea levels, storms, water shortages and droughts. By reviewing this classification of sectors and impacts, in the characteristics of European cities, they manage to build interpretive scenarios that recognize the strengths and weaknesses of the experiences when seeking to minimize the vulnerability of cities to climate change.

Given that the vulnerability analysis was carried out in the city of Bucaramanga, its position and geographical configuration conditioned the list of variables, characteristics or sectors identified. Of the aspects that the Municipal Mayor's Office highlights about the territory, located in the Colombian Andes, Bucaramanga is located on a sloping terrace of the Eastern Cordillera, its soils can be divided into two groups: the first, having no danger of erosion, are conducive to the cultivation of a wide variety of products and use for livestock farming. The other type of soil has a high erosive potential; For this reason, it has low fertility and a surface fertility layer, in some situations almost zero, the municipal area is 165 square kilometers, its height above sea level is 959m and its thermal floors are distributed in: warm 55 kilometers squares: medium 100 square kilometers and cold 10 square kilometers. Its average temperature is 23°C and its average annual precipitation is 1,041 mm.

The position and configuration of the city conditions the variables and sectors taken as reference, given that they can be regrouped in spaces present in the city or discount those that do not apply to the territory studied. With this, the analysis sectors to be worked on have been determined to be four: extreme temperatures or heat waves; roads, mobility and vehicle fleet, solid waste management and existence of public spaces and green areas.

For its recognition, its existence in each commune was analyzed, its level of impact due to exposure, sensitivity and information and capacity to adapt to climate change was catalogued, with which it was granted a condition of vulnerability according to the criteria applied corresponding to one of the four stipulated levels and with this, establish the level of vulnerability of the commune, to consolidate in a qualitative average exercise, the condition or level of vulnerability of the city. This process and its global structure were taken from the model exercise developed in commune 13 of the city (Amaya y otros, 2019).

As a criterion to classify the level of vulnerability with which a territory is affected, the criteria proposed by the Green Cities Index of the multinational SIEMENS (EIU, 2010)and adapted by Amaya (2019) in its methodological proposal were followed.

Table 3. Description of vulnerability levels

Vulnerability Levels	Definition
High	The variable must be intervened as soon as possible, it generates damage.
Half	Intervention is required to strengthen the processes to avoid significant deterioration.
Low	There is no strong deterioration but control measures must be taken.
Not Altered	It is not vulnerable.

Source: taken from (Amaya y otros, 2019).

The vulnerability criteria were analyzed in each commune, determining an aspect of recognition or existence of each of the four variables, according to exposure, sensitivity and information and adaptation capacity.

Table 4. Vulnerability Criteria for the variable Extreme temperatures or heat waves

Extreme temperatures or heat waves			
Vulnerability Levels	Exposure	Sensitivity	Information and adaptability
High	There are records in the last five years of effects and damages, associated with the urban development of the city and due to strong periods of heat, planned or not, climatically.	There are effects such as population displacement, loss of crops, and land uselessness as a result of high levels of heat, above the historical average for the area.	Information and knowledge is insufficient, there are no clear strategies on responses to incidents of hot seasons
Half	The construction development of the city has modified average levels of ambient temperature in the city, changes in atmospheric humidity and the incidence of sunlight.	There are records of special effects on the population, in the summer season, skin damage, deterioration of coverage, impact on flora or fauna.	There is evidence of campaigns and information processes on climate change and possible effects or changes, but there are no actions in processes on the population.
Low	In historical summer times, there are higher temperature levels and although it requires changes in urban life, the city dynamics are maintained.	Minimal alterations to population, green areas or flora and fauna as a result of changes in ambient temperature.	After citizen training campaigns and development of public infrastructure, the population achieves understanding and prevention of the effects of climate change
Not vulnerable	There are heat islands, but neither the population nor the urban system in general has been affected.	There is an increase in temperature but it does not generate alterations in the community	Information exists and all actions have been taken to mitigate the risk.

Source: Author's own elaboration.

Table 5. Vulnerability Criteria for the variable roads, mobility and vehicle fleet

Roads, mobility and vehicle fleet			
Vulnerability Levels	Exposure	Sensitivity	Information and adaptability
High	Road deterioration, poor ventilation due to urban construction conditions, high levels of vehicle congestion and level of air quality with records of high exposure to automotive CO2 and medical evidence of effects on human health	Air quality is poor, high concentration of pollutants identified from automotive sources, mobile sources, has high levels of impact on people's health and deterioration of green areas.	There are no processes to regulate automobile traffic and there is no evidence of organization on the roads, proliferating free traffic, impacting the entire city.
Half	High levels of vehicle congestion, with regulatory measures to reduce effects on the population and the environment in general	Air quality levels show deterioration as a result of the road system and vehicle fleet; measures to regulate vehicular traffic are identified.	Although there are citizen education campaigns to improve the behavior of road actors, no changes are evident in road axes and vehicle ownership and use.
Low	Application of regulations to the road system, orderly road flows, good levels of urban air quality	Air quality levels are acceptable, low concentration of pollutants, good conditions of urban flora and fauna.	Road development improves as it is renewed and vehicular traffic complies with restrictive regulatory measures for environmental impact.
Not vulnerable	The road system and vehicular traffic do not generate sensitive environmental impacts on the population.	Air quality is satisfactory, pollutants dissipate easily	The road axes promote agile and sustainable road circulation and the community minimizes its impact on the urban environment in general.

Source: Author's own elaboration.

Table 6. Vulnerability Criteria for the solid waste management variable

Solid waste management			
Vulnerability Levels	Exposure	Sensitivity	Information and adaptability

High	Epidemics occur due to the proliferation of vectors and bad odors, resulting from inadequate accumulation of waste. Poor waste disposal is evident throughout the city	There is a total absence of citizen education in waste management, there are high levels of disease vectors and proliferation of polluting sources.	There is no citizen education process regarding the selection and disposal of urban solid waste. There is no urban infrastructure for temporary disposal and collection and final dissipation does not respond to demand.
Half	There are pollution problems in public spaces and green areas, due to the low management capacity of the municipal authority, areas with waste disposal problems have been identified.	Landscape damage in specific sectors, with lack of waste management, sources of waste that generate disease vectors	There are government campaigns and investments for the infrastructure and waste management process but the citizen response is low, maintaining sources of waste proliferation and disease vectors.
Low	The authorities apply permanent comprehensive solid waste management campaigns, having small sectors with associated problems; the impact points for urban solid waste are minimal.	There is citizen government coordination for the temporary disposal and adequate transportation of solid waste, controlling difficult collection points through campaigns and continuous attention.	The government-citizen articulation achieves the selection, delivery and organized transportation of waste. Contamination points remain but they are intervened periodically
Not vulnerable	The collection routes are efficient, the waste management process responds efficiently in all stages of collection, transportation and disposal.	The comprehensive solid waste management system works correctly	There is the information and training required for the correct functioning of the PGIRS.

Source: Author's own elaboration.

Table 7. Vulnerability Criteria for Existence for the variable public spaces and green areas

Existence of public spaces and green areas.			
Vulnerability Levels	Exposure	Sensitivity	Information and adaptability
High	There are no consolidated green areas in the city, there is evidence of deterioration of parks and common areas, minimal or no urban fauna. There is no urban road coverage	Environmental damage has deteriorated urban forest or forest cover and public parks show denuded soils with minimal existence of green surfaces.	There is no green urban infrastructure that balances city construction. The urban infrastructure does not have a green component that articulates with the dynamics of urban life.
Half	There are green areas in the city, forest cover is evident in development and intervention is required for its consolidation. Public spaces require care to consolidate trees	Parks, green areas, and road axes show affected forest cover, with evidence of interventions for their recovery. Protection and conservation campaigns are being carried out for spaces with urban flora and fauna.	Improvement intervention actions on parks and green areas are evident, the forest cover has protection and conservation measures and the community has access to parks and green areas, although with a low level of orientation and training
Low	The green areas have significant forest coverage, the road axes have gallery trees, public spaces are cared for and protected, urban repopulation of flora and fauna is encouraged.	Public space and green areas with consolidated forest cover, which require permanent protection and conservation, restoration or revegetation processes are carried out, urban fauna and flora spaces are protected.	Urban parks and green areas have consolidated coverage, the community recognizes the importance of urban green infrastructure and enjoys natural areas as a balance of urban life
Not vulnerable	The urban green areas are articulated in the development of the city and the road axes have consolidated forest cover, the articulation of the city with green covers is evident and the spaces of urban flora and fauna are respected.	The green areas, public parks, road axes, have permanent protection, conservation and care, the community has access to recreational activities and passive green spaces and the urban flora and fauna has delimited and protected spaces.	The city has green areas, public parks, road axes with consolidated forest components as structural axes of its formation where urban flora and fauna are protected and consolidated. There is an urban ecological structure known and used by the community.

Source: Author's own elaboration.

Once the interpretation criteria were determined, the identification of the information by communes allowed us to build a general city scenario, based on the vulnerability conditions of its entire territory. Although the information is not homogeneous throughout the city, there are many coincident aspects that confirm the city's identity. A good level of citizen identity is recognized that seeks to take care of the city, green coverage, parks are spaces of special recognition where the community still sees the social meeting point, the average size of the city fosters greater urban integrity; Public services are recognized as having good coverage and a good level, solid waste management seeks to cover the city, although sustained population growth increases the delivery of waste for collection, transportation and disposal, no one denies the difficulties associated with landfilling. health of the city and the need for its recomposition and relocation; In solid waste, market places are a generating focus that requires special attention to achieve better waste management conditions.

Even so, there are also negative aspects that any commune identifies as environmental problems of the city, the high levels of vehicular congestion, explained by the permanent growth of the automobile fleet, without renewal of the automobiles that deteriorate and the perpetuity of the road axes since their construction, without being resized to the current population levels and urban structure of Bucaramanga; Added to this is that the mass transportation system fails to cover the entire city, so it does not cover the travel needs of the residents, generating informal transportation, usually in vehicles without technical control in their operation, or the increase in micro trips. between areas of the city, extending routes that before the system were made in a single means of mobility.

The main aspects associated with the variables selected to interpret vulnerability to climate change are:

Environmental temperature

The temperature in the department of Santander has increased in recent years, which in turn reflects a direct relationship with the deterioration of ecosystems, as stated by the Ministry of the Environment who adds that the Colombian territory will increase 2.3 degrees Celsius due to to climate change.

The city does not present significant changes in the hot seasons, although this increase in temperature is recognized, but this does not mean that particular situations derived from atmospheric conditions or alterations are identified. In general, data from the Industrial University of Santander and IDEAM demonstrate that the city's climatic averages have behaviors that, although they can generate alterations in the dynamics of life, do not go beyond historical parameters.

" *The average annual total rainfall is 1303 mm. During the year the rains are distributed in two dry seasons and two rainy seasons. The driest months are December, January and February, and to a lesser extent, June, July and August. The rainy seasons extend from March to May and from September to November. In the dry months at the beginning of the year, it rains around 10 days/month; In the months with the highest rainfall, as well as in the dry season in the middle of the year, it can rain 17 to 19 days/month. The average temperature is 22.6 °C. At midday the average maximum temperature ranges around 28ºC. In the early morning the minimum temperature is between 18 and 19ºC. The sun shines a little less than 4 hours a day in the rainy months, but in the dry months, sunshine registers between 5 and 6 hours a day. The relative humidity of the air is greater than 80% on average and in rainy seasons it reaches values above 84%* . (IDEAM I. d., 2015)

Roads, mobility and vehicle fleet

This is a neural issue of Bucaramanga, the development of the city has as an ineffable limitation its geological-geographical position, the city was settled on an alluvial fan adjacent to a seismic knot and bordered by significant ecological components: the eastern hills, paramo area and the Rio Frio basin. Due to these conditions, the city must rebuild, rethink, relocate, redistribute itself and this has had the least repowered element, the road axes, given that the streets, lanes and avenues have remained the same for more than 30 years, but To date, the city has almost quintupled its resident population and has transformed its city structure, from large family mansions to large multi-family vertical constructions.

The road load and motor traffic identified coincide with aspects expressed by the Metropolitan Area of Bucaramanga, AMB, in the Mobility Master Plan 2011-2030, of which aspects such as: vehicular congestion as the main problem perceived by citizens are highlighted. from the city of Bucaramanga. Fundamentally caused by Bucaramanga being the social, administrative and commercial center of the metropolitan area and the department. To this we must add the very high motorization rates of the last decade, and highlight the participation of motorcycles. Areas such as Centro, Cabecera, Ciudadela Real de Minas, San Francisco and La Concordia, which absorb about 40% of the municipality's trips (around 55,000 daily trips), understood to be the main residential and commercial sectors of the city. As notable data, the AMB highlights that it is estimated that the city's vehicle fleet increases annually by 6,250 vehicles (AMB, 2011), the author's fleet projections estimate close to 130 vehicles per thousand inhabitants in the year 2025.

Solid waste management

Solid Waste in general in Bucaramanga has a medium level of satisfaction, in the sense that the vast majority of the city has a system of collection routes that serve the city's production. The service providers serve all sectors and achieve the collection to be transported to the Carrasco landfill. In the sectors with the highest residential load such as the capital commune, real de minas, mutis, occidental, Provenza, the delivery of solid waste is organized at provisional or door-to-door delivery points by residential conglomerates. In city areas where residential complexes do not predominate, there is a greater presence of waste on public roads, although it is collected regularly by service providers.

A component of special attention is the points of sale of agricultural production, until 10 years ago, in the so-called market houses of the municipality, but in the last five years they have multiplied as commercial points throughout the city. In any case, they are points of high production of organic waste, which historically have been a point of contamination of soil, air and water and generation of vectors affecting the health of nearby residents. It is noteworthy that these impacts have been handled favorably, especially the market houses have been refocused and today much more orderly and clean environments are seen, without the indiscriminate presence of solid waste.

A highly negative point for the city is the management of the regional sanitary landfill, the carrasco, which has already reached its useful life and its technical storage capacity has been filled. Political irresponsibility and the need to maintain public health have forced the last ten years to extend the operation of the carrasco, although socially and environmentally it should already be closed and the city have advanced towards modern, technical and sustainable systems for using waste, leaving disposal by burial, as the last possible element and with minimum amounts to be allocated. This difficulty with the final disposal site is a situation that affects the entire city and although in the urban environment it does not show in a significantly impactful way, disorder in solid waste, the inefficiency of the final link in the management process weakens the capacity of city to contain the incidences of its solid waste in the face of climate change.

Existence of public spaces and green areas

A historical characteristic of Bucaramanga is the significance of green areas in its urban infrastructure; in the 80s and 90s it received recognition as a city of parks, due to the good relationship between inhabitants and existing parks.

Today the city's parks constitute the green element that provides open spaces, with natural components, to balance the growth of urban construction in the territory. In the vast majority of parks, the miscellaneous tree component stands out, perhaps the most outstanding is the San Pio park, in the main commune, which stands out for its solid forest cover and its constitution of a natural microspace within the urban dynamics. Other parks, spread throughout the communes, San Francisco in the San Francisco commune; Las Cigarras, in the Real de Minas commune; Mutis in the Mutis Commune, La Concordia, Antonia Santos in the La Concordia Commune; García Rovira, Centenario, Santander in the central commune and 6 other municipal parks, constitute the main relicts of green areas, connected by road axes with tree cover, contributing to the ecological structure and the dynamics of urban flora and fauna, next to the forest consolidated area of the eastern hills of the city and the slope ecosystem of the western escarpment of the colluvium, constitute the setting for the generation of ecosystem services that the city directly takes from its nature.

With the general information collected in the communes, the criteria were organized in an analysis matrix, to estimate a level of vulnerability, at the intersection of each chosen variable with the determined criteria, following the criteria described in tables 1 to 5 above. , and then, by average prevalence, assume a level of vulnerability with qualitative arguments for each commune, from which, consolidate a general city average for its level of vulnerability.

Table 8. Vulnerability by communes, according to the relationship between variables and criteria

Communes	Vulnerability Criteria			Vulnerability
	Exposure	Sensitivity	Information and Capacity	
North	Half	Half	Half	Half
Northeast	Half	Half	Half	Half
San Francisco	Half	Low	Half	Half
Western	Half	Low	high	Half

García Rovira (Central-Eastern)	Half	Low	high	Half
La Concordia (South-Central)	Half	Low	high	Half
The Citadel (Central-Western)	Half	Half	Half	Half
Southwest	Half	Half	Half	Half
La Pedregosa (South-Eastern)	Half	Half	high	high
Provence (South-West)	Half	Half	Half	Half
South Commune	Half	Half	high	high
Head of the plain (Eastern)	Half	Half	Low	Half
Eastern Center	Half	Half	high	high
Morrorrico (Northeast)	Half	Half	Half	Half
Center	Half	Half	Half	Half
Cacique Lakes (South-Eastern)	Half	Low	Low	Low
Mutis (Western)	Half	Low	Low	Low

Source: Author's own elaboration.

The analysis of the existence and representativeness of the four selected variables, contrasted with the criteria that allow qualifying the territory's response to the incidence of climate change, made it possible to identify a general level in each commune. With this information and by representativeness, a general level of vulnerability of the city to climate change was identified, estimating that the city can be recognized as having a MEDIUM level of VULNERBILITY after equipping the conditions of the different spaces analyzed.

Although Communes such as Algos del Cacique and Mutis have a low level of vulnerability, the particular conditions of these communes are not repeated throughout the city, nor are they the strongest in the entire city. Aspects such as mobility, which in Cacique stands out for being mostly through private vehicles, with a low presence of heavy transportation, make the atmospheric conditions more favorable to the health of the inhabitants in this area, in addition to have abundant green areas, both public and within the different residential complexes, and citizen order in the delivery of solid waste. For its part, in the commune of Mutis, although there is a greater presence of public transportation, in general there are agile conditions for the movement of the population and permanent flows that encourage the circulation of polluting gases in the area, in addition, their route. due to the escarpment of the city and micro-basins of urban streams, it is a favorable condition to maintain good atmospheric conditions, this marks an important natural coverage, although it is affected by poor citizen management of solid waste, although not generalized but in localized points of the commune. In the other sense, the communities that were found to have high vulnerability, in aspects such as mobility, are affected by a vehicular overload, between public and private, high levels of atmospheric gases derived from automobiles, few green areas in good condition and proliferation of waste. solids in their environment.

Assuming a condition of Medium Vulnerability for the entire city, recognizes the strengths and weaknesses of the city, which allows the space to continue the processes that improve the urban environment and thereby intervene in the aspects of difficulty that negatively affect the quality of life of the city.

From what is presented by ECLAC (Sánchez, 2013), it is possible to interpret that the cities of Latin America have very similar aspects in their installed capacities, in their social structure and in their availability of resources. Latin America stands out in the urban area for being population centers that start from their historic center and with the demographic increase, they cover peripheral lands that are gradually incorporated into the urban dynamics and becoming new centralities for the permanent growth of the cities. This means that aspects such as mobility have to be dynamic to adapt to urban growth. There will be periods of inefficiency, but proper planning of the territories will include ensuring that the possible mobility mechanisms grow appropriately with the city. Along with the process of demographic growth, the generation of solid waste is an inevitable consequence. In Latin America, in a country like Brazil, there are very successful experiences in which it is possible to minimize the effects of waste and, on the contrary, convert it into urban potential, for example. , the socioeconomic dynamics in Sao Paulo, after achieving comprehensive welfare processes with solid waste (Margulis, 2016).

As presented by Margulis (2016), city problems such as solid waste, green areas, citizen education, require permanent intervention from the authorities to support them as the main aspects of help to minimize potential effects due to climate change. In the case of Bucaramanga, the exposure of climatic incidents cannot be modified, if the aspects of urban life that do not increase the difficulties in controlling climate changes can be strengthened.

The management of solid waste is a permanent need for control, requiring rethinking from the very generation and selection at the source, there the citizen training processes must remain in force so that the first link of solid waste minimizes the negative influence of these on the urban environment, from this, the collection, transportation and final disposal process has to be strengthened by having technical waste management. Bucaramanga requires that its waste management incorporate new technologies and be regulated by phase, to promote the functional recovery of waste and the minimum space requirement for the confinement of what is not usable.

Three phases must be consolidated, one citizen education for management at the source and final disposal, supported by incentives, but also sanctions, which show citizens' duties and rights, for clear compliance with aspects such as the selection, disposal and delivery of waste, to the collection and transportation processes. The second phase, collection and transportation, must be under full government supervision and be an example of action for the citizen response. This stage must contemplate efficient routes, in route, time and technologies, as well as what the citizen shows that the process contributes to their living environment, beyond just removing waste, but also participates in a timely manner in their needs. Phase three, which is not visible to the common citizen, must be handled in two minimum moments, one, delivery of the initial collection to collection, selection and transfer stations, from which usable waste is identified and promote their reincorporation into productive systems, after which, a second phase of transportation and final disposal where they are documented as a minimum load of waste that is disposed of in technical spaces, which generates natural destruction and reconversion, without affecting the associated living environment.

In mobility, the cities of Latin America have sought to streamline it through the mechanisms of mass transportation systems, seeking the multimodal integration of forms of urban mobility, focused on minimizing costs, impacts and social times, maintaining the connectivity of living areas with those of marketing and citizen coexistence. The approach of Mass Transportation systems, promoted by the Inter-American Development Bank, through its Sustainable Cities strategy (BID, Banco Interamericano de Desarrollo., 2106)is an example of this, Bucaramanga developed its mass system since 2007 and came into operation since 2011 (AMB, 2011), covering the main axes of the city, although not all areas of the city, this caused the mobility of the city to become lethargic and overcrowded, the population looked for alternative means that would allow them to maintain their movements, although outside the mass transportation system and detracting from the benefits with that the system was designed for the city. Today, Bucaramanga generally has multiple sources of traffic congestion, which generates an alteration in air quality, given the increase in greenhouse gases, although they are not permanent over time, although in some seasons and areas of the city it has There have been levels of deterioration in the air condition, which has an impact on the health of the inhabitants, the flora and fauna of the city. Based on the ideas of Sánchez (2013), (Sánchez, 2013)urban mobility should be a fundamental component of the capacity for adaptation and mitigation of climate change in cities. The increase in the number of vehicles is inevitable, but this must be managed by local administrations. with restrictive measures or with incentives for better road behavior. In that sense, Bucaramanga, like all cities in Colombia, has implemented the Pico y Placa restrictive measure, which seeks to reduce, by at least 40%, the number of private vehicles in circulation per day and should promote the use of transportation. massive, unfortunately for Bucaramanga, this has not been presented like this, mass transportation has not managed to be the axis of city mobility and the roads have become small due to the strong demand for growing automobile fleet. In addition, Bucaramanga, as a pole of departmental and regional development, is the attractive urban environment of the other metropolitan municipalities: Floridablanca, Girón, Piedecuesta, Lebrija, are the origin of a large number of vehicular trips, due to their dormitory vocation, which unload a greater number of vehicles. in the city. In addition to Pico and Placa, a reconstruction of the road actor is required, which is why it is imperative to execute strong citizen awareness campaigns on the sustainable use of public roads, responsible behavior in vehicle ownership and use, tolerance when sharing public space and of individual contributions for the regulation and improvement of vehicular flow in the city. Measures such as restricted traffic zones, satellite parking and congestion flows should not be necessary in Bucaramanga, but given the unstoppable increase in the number of vehicles, a greater sectorized organization of vehicular traffic and a resizing of the mass transportation system are required for its real protagonism. in city mobility.

A significant component of the response to climate change are green spaces and public areas in urban environments. Without a doubt, forest cover in cities is the main mechanism to reduce incidents of heat, strong exposure to sunlight, capture and dispersion. of greenhouse gases, the existence of flora and fauna and in general, a component that balances the constructive development of a city, through spaces where the inhabitant comes into contact with natural elements. For both CEPLA, through Galindo et.al (2015) (Galindo y otros, 2015)and the experiences compiled by the European Environment Agency (AEMA, 2015), natural coverage in cities is the main component of the response to climate change, therefore administrations must establish and strengthen the measures that maintain existing natural spaces and promote in new urban developments the imperative incorporation of natural components, not only for compensation, which is usually done extraterritorially, but within urban designs and distribution of territories, forcing the incorporation In the growing city, green areas as an urban structural element. The public parks of Bucaramanga are the key relics for the sustainability of the ecological structure of the city, the forest covers in them must be strengthened, improve the green coverage and promote the existence of flora habitat for urban fauna. The San Pio park in the Cabecera commune is the best ecologically consolidated park; in all the communes there are historically and socially significant parks, which should increase their occupation due to vegetation cover. The ecological structure of the city (Bucaramanga, 2012), between the eastern hills and the western escarpment, the parks are the links, but these two extreme points are required to be connected by corridors that promote the mobility of fauna and flora and consequently repopulate green spaces. open in Bucaramanga. The main road axes have consolidated trees, which are a necessary element for the thermal regulation of the urban environment, but nevertheless their entire environment must be strengthened, not leaving the tree alone but seeking green coverage and bushes that are repositioned to consolidate. According to the World Health Organization (WHO), the ideal minimum urban public space index should be greater than 9m ^{2} per inhabitant. The authorities of Bucaramanga show numbers very close to this, when they incorporate the large natural extension that constitutes the eastern hills, but If these are discounted and only take the green spaces within the urban development, the public space index in Bucaramanga does not exceed 3.5 m ^{2} per inhabitant, a fairly low value and which denotes the imperative need to increase this proportion, in the face of adaptation and mitigation of climate change. An important contribution to the green areas and public space in the city are protected nature reserves through the figure of ecological parks, Parque la Flora, the Quebrada La Iglesia linear park and the Carlos Virviezcas zonal park (Bucaramanga, 2012), should be a space to repeat in the entire city, where urban development is harmonized with the existence of attractive natural covers of Flora, generating shade, consequently climate regulation and carbon sinks, promoting the mitigation of greenhouse gases.

CONTRIBUTION TO THE SDGS AND SUSTAINABILITY

The sustainable development objectives, although they are not a mandate for compliance, do constitute a common vision towards common well-being, their importance is positioned to the extent that environmental, economic and social actions achieve more equitable realities for all living beings and to the extent that we all integrate ourselves into them and ensure that our actions are linked to the goals.

Here an important space for reflection opens in which all social actors find a space to participate in improving the quality of life of everyone in the city. For the ordinary citizen, having a balanced living environment that ensures access and satisfaction that allows him to care for himself and be cared for with opportunity, to the ruler, because it opens up the spectrum of focusing efforts on sensitive aspects of the community, which at the same time Searching for intervention measures on a front can generate indiscriminate benefits for all inhabitants, in addition the irradiation of public investment will be more easily evident by the inhabitants when visualizing the reconstruction of the city in aspects of everyone's problems. And for the businessman, who provides the economic structure that moves the citizens, the economy, the productivity, the profitability of the city, recognizing climate change opens up space for social positioning, in which he achieves a double action, which contributes to the common well-being, but that strengthens the economic support of its activity: in general, become an active part of the initiatives to intervene in problems derived from climate change, where the community in general identifies the commercial agent not only as the one that provides a resource or service, but as cares about its environment and seeks its well-being with reach for everyone; Specifically, and it may be an aspect of greater impact, gradually transforming its activity to avoid the generation of conditions that impact the main problems of the city in the face of climate change, reduction of solid waste, sustainable mobility, environmentally friendly products and environmental education. to transform social consciousness from its employees to the entire community.

This second moment, although it seems more complicated, can be the most favorable for the company, transforming its production system, reducing the consumption of natural resources, substituting impactful inputs and evidencing environmental values in its products, has a significant impact on the communities; Using means of mobility that are friendly to everyone, with minimal emissions, noiselessness and efficiency in movement leaves the inhabitant with the impression of true contribution to the well-being of all, without stopping their productive systems. And developing a training process for employees, which includes environmental commitment, becomes a multiplier of messages that manage to extend the business commitment, beyond the physical limits of the company.

ECLAC has established that "Responding to the challenge of climate change in Latin America and the Caribbean represents a financial, economic, social, cultural, distribution and innovation effort, but it also provides an opportunity for the region to move towards more sustainable development and inclusive" (Bárcena y otros, 2020)with which it makes a commitment to the comprehensiveness of efforts to respond to climate change, in addition to giving a prominent role to the government and businessmen, to appropriate initiatives, position themselves socially and transform their productive systems, leading communities in changing customs and in immediate and prospective responses to climate change.

Through multiple channels, the UN expresses the scope of the consequences of climate change, for companies it proposes how climate changes will have a direct impact, forcing changes from their infrastructure to their priorities and forms of investments. Along these lines, the approach articulated within the SDG has the particularity of not having exclusive results; a single action can help enhance several objectives and within them several goals to be achieved. Actions focused on climate change have multiple repercussions and in the SDGs their scope supports well-being in many aspects of the proposals from this global vision.

Climate change affects, among many other aspects: the capacity of soils for agricultural production, air quality, the availability of water for livestock production and for human consumption, availability of resources for all productive systems, balance of terrestrial ecosystems, dynamics planetary climates, development of cities and infrastructure of all kinds. Due to the above, the linking of actions for adaptation and mitigation to climate change could be taken from any of the SDGs and project contributions for the others, but in the search to facilitate their application, the most appropriate thing is to focus the actions from SDG13 Action for Climate, to then project benefits radiated to other SDGs, fundamentally and through the focus of this chapter, to SDG11 Sustainable and Resilient Cities, also to SDG15 Protection of Terrestrial Ecosystems.

From SDG13, atmospheric and climatic problems go through the alteration of the earth's greenhouse effect, its decompensation due to the emission of gases from anthropogenic activities, must be the subject of intervention of environmental processes, climate change as the main effect, must be intervened from citizen training, through technical processes and relying on government policies and strategies that help improve the quality of life of all inhabitants. The vulnerability analysis of the territories gathers basic information for these processes, given that it allows us to identify the aspects in which society most recent climatic variations, identify the activities that are contributing to this deterioration and recognize the focuses of citizen education in which work, to transform behaviors that impact mitigation, adaptation and better living conditions.

The goals proposed from SDG13 require the Vulnerability process as the basis of information for decision-making.

Strengthening resilience and the capacity to adapt to risks starts from knowing the internal factors that to a greater or lesser extent will be affected by the external agent that climate change becomes. Resilience requires citizen education processes to respond to the situations and the capacity to adapt leads to actions that allow living conditions to be maintained and not increase deterioration due to the climate. The formulation and implementation of territorial policies, strategies and plans is a key tool for the social acceptance of actions leading to the management of climate change, given that they provide legal and social compliance guidelines from which responsibilities and processes that provide true well-being to society are recognized. And Improving education, awareness and human and institutional capacity regarding climate change mitigation will be a result of integrated management between administration and society, where citizens recognize in their actions the positive and negative consequences on climate change and Find support in the institutional guidelines to transform your processes, life habits and strengthen coexistence spaces.

The medium vulnerability condition assumed for the city of Bucaramanga creates spaces for adaptation and mitigation processes to focus on developing aspects included in SDG13. It is necessary to start with citizen education processes, focused on three aspects: responsible mobility, citizen management of solid waste and use and care of green areas and public spaces. The city community must have permanent processes that remind us of the importance of acting on these environmental fronts, highlighting the common benefits that can be achieved.

Here the business and productive sector of the city is called to lead processes of social transformation, in which its product transmits not only the commercial purpose for which it was conceived, but also assumes a transformative message in the face of climate change. For the UN, local actions must generate global benefits. Under this approach, citizens, businessmen and the government must take their own actions that permeate the well-being of all and that respond to the challenges that climate change affects the city. For Juan Carlos Moreno Brid from ECLAC (2020), the business sector can have great benefits if it assumes leadership in the face of climate change, estimates for Latin America bet that the cost of climate change can be between 1.5% and 5% of the regional GDP in 2050, but if actions and investments are assumed to transform and modernize the productive structures that represent current adaptation costs, they may be less than 0.5% of the 2020 GDP of the region.

These aspects seem not linked to business activity, but when climate change is identified as an externality of economic growth, as proposed by Moreno Brid (2020), companies can find important opportunities for social positioning and commercial growth, if they lead a transition. towards the green economy, they mark the reduction of poverty and minimizing their carbon emissions.

From the city analysis, actions can be pointed out to reduce vehicular chaos and consequently improve air quality conditions in the city, which will be reflected in strong urban green coverage and better health conditions of the inhabitants. Adequate management of solid waste that prevents the proliferation of health vectors in sources of undue accumulation or its disposal in unforeseen ways, times or spaces, which deteriorate the general landscape of the city. And construction of identity of the inhabitants towards their city in which green spaces and public spaces are recognized as infrastructure that improves their quality of life, in which everyone can enjoy and must exercise responsible actions that build and improve the spaces already provided..

With permanent processes of environmental education, the municipal administration must lead management actions that efficiently complement community contributions, that is, provide green areas that are cared for and adapted for the enjoyment and contemplation of residents, where passive recreational interaction activities can be carried out. with green environments and the strengthening of the health of all people is recognized, as well as the implementation of efficient mechanisms in waste collection and the implementation of technologies that reflect continued well-being based on the initial contribution of the citizen, communicating the minimization of problems social problems due to the permanent and inevitable generation of solid waste.

CONCLUSIONS AND RECOMMENDATIONS

The problems of city planning and development coincide with the problems seen from climate change, solid waste management and vehicular chaos due to old roads, they are the aspects that hit the hardest from the urban infrastructure and hinder the vulnerability conditions of the city. city. Although in the vast majority of sectors, these conflicts have been managed, their regulation must be intensified to prevent vulnerable conditions from continuing to deteriorate. For Bucaramanga, it is imperative to solve the problem of solid waste management, thereby promoting the health of the inhabitants and also dynamizing vehicular traffic, managing the renewal of road axes, to support the greater demand for automobile fleet and improve the control infrastructure. road to streamline traffic on main travel axes and consequently optimize travel times within the urban environment

Climate change is a factor affecting all living environments, the urban environment makes these effects strongly evident by agglomerating multiple activities and lifestyles in a restricted territory, which is why acting on people must be a more solid support to transform the city. and improve their conditions of vulnerability, if people do not transform their customs, attitudes and behaviors, their territory of life continues its degradation. The most important bet must be the change in attitudes which leads society to carry out territorial changes that achieve quality of life after the balance of man and nature.

It seems that the urban aspects of climate change do not have a direct relationship with the business and commercial reality of the city, but it is undeniable that the deteriorated factors of the city affect the levels of commercial exchanges. Vehicular chaos determines mobility limitations, resulting in rethought and relocated movements in the pursuit of goods and services; Solid waste affects the landscape and health of the environment, therefore regulating citizen access by city sectors; The loss of identity and belonging of the citizen to their environment, leads to ignorance of the productive activities of the territory, all aspects that break commercial dynamics and consequently the business life of the city. The company participates in a change in these realities associated with climate change, acting from the inside out. Changing and improving their production systems, reducing their generation of solid waste using more reusable resources, in situ and in the useful life chain of their products, promoting better trained and trained employees not only technically but also humanly, are actions that They lead to opening new social spaces, where the renewed awareness of the environment seeks products that also care about that environment. Although this recapture of buyers is not immediate, it can be a result that prospectively leads to more economically favorable scenarios.

REFERENCES

EEA, A.E. (2015). *Climate Change and Cities.* Madrid, Spain: European Environment Agency.

European Environment Agency, EEA. (2018). *10 Case Studies How Europe is adapting to climate change.* https://climate-adapt.eea.europa.eu/about/climate-adapt-10-case-studies-online.pdf

MAYOR OF IBAGUE. (2016). *alcaldiadeibague.gov.co.* http://www.alcaldiadeibague.gov.co/portal/admin/archivos/publicaciones/2016/14024-PLA-20160502.pdf

Ibague Mayor's Office. (2016). *alcaldiadeibague.gov.co.* http://www.alcaldiadeibague.gov.co/portal/admin/archivos/publicaciones/2016/14024-PLA-20160502.pdf

AMAYA, CC, & HERNANDEZ, CC (04/15/2018). *Pascualbravo.edu.co/cintexpb* . Pascualbravo.edu.co: http://www.pascualbravo.edu.co:5056/cintexpb/index.php/cintex/article/view/301

Amaya, CC, Tavera, CN, Avila, AM, Alvarado, PJ, & Vesga, Á. P. (2019). Methodological Proposal for Evaluation of the Level of Vulnerability to Climate Change in Urban Environments. Commune 13, Bucaramanga, Santander. *17th LACCEI International Multi-Conference for Engineering, Education, and Technology* (pp. 1-10). Boca Raton - USA: LACCEI.

AMB, A.M. (2011). *Mobility Master Plan for Bucaramanga 2011-2030.* Bucaramanga: AMB.

Arboleda, A. (2008). Manual for the evaluation of environmental impacts of projects, works or activities., (pp. 75 - 86). Medellin.

Barcelano, A. (2020). *Barcelona for Climate, Ecology, Urban Planning, Infrastructure and Mobility.* https://www.barcelona.cat: https://www.barcelona.cat/barcelona-pel-clima/es/barcelona-responde/acciones-concretas

Bárcena, IA, Samaniego, J., Peres, W., & Alatorre, JE (2020). *The emergency of climate change in Latin America and the Caribbean: do we continue waiting for the catastrophe or do we take action?* Washington DC: United Nations - ECLAC. LC/PUB.2019/23-P. ISBN 9789211220315.

BARTON, JR (2009). Adaptation to climate change in city-region planning. *Magazine of Geography Norte Grande, 43* (1), 5-30.

Bauni, V. A. (2017). Wildlife mortality due to road kill in the Atlantic forest of Alto Paraná, Argentina. *Ecosystems Magazine* , 54-56.

IDB, BI (2016). *Evaluation of the IDB Emerging and Sustainable Cities Initiative.* Washington, DC: IDB.

IDB, Inter-American Development Bank. . (2106). *Methodological guide of the Emerging and Sustainable Cities Program.* Washington DC USA: IDB.

Bogota. (2015). *District plan for adaptation and mitigation to climate variability and change.* Bogota.

Bonds, B. 2. (2001). Wildlife habitat mitigation. *In: Wildlife and highways:seeking solutions to an ecological and socio-economic dilemma. 7th Annual Meeting of The Wildlife Society. Nashville, Tennessee* , 70-72.

Breil, M., Downing, C., Kazmierczak, A., Mäkinen, K., & Romanovska, L. (2018). *Social vulnerability to climate change in European cities – state of play in policy and practice.* Bologna, Italy. https://doi.org/10.25424/CMCC/SOCVUL_EUROPCITIES: European Topic Center on Climate Change impacts, Vulnerability and Adaptation (ETC/CCA).

Bucaramanga, AM (2012). *Territoiral Planning Plan 2012-2027.* Bucaramanga: Mayor of Bucaramanga, http://www.concejodebucaramanga.gov.co/pot-2012-2027/tomo01.pdf.

Buenos Aires, G. d. (2017). *Provincial Report, Adaptation of the Sustainable Development Goals in the city of Buenos Aires.* Buenos Aires, Argentina: Government of the city of Buenos Aires.

Cali, AM (2020). *Municipal plan for adaptation and mitigation to climate change.* Cali. Cauca Valley.

Canales-Delgadillo, JP-C.-J.-P.-P.-M. (2020). Traffic deaths on the Gulf of Mexico coastal highway: how many and which species of wildlife are being lost? *Mexican Biodiversity Magazine* , 91.

Castillo-R, JC-M.-G. (2015). Wildlife mortality due to vehicle collision in a sector of the Pan-American Highway between Popayán and Patía. *Scientific Bulletin. Museum Center. Museum of Natural History,* , 207 -219.

Chavarro, PM, Garcia, GA, Garcia, PJ, Pabón, JD, Prieto, RA, & Ulloa, CA (2013). *Preparing for the future, threats, risks, vulnerability and adaptation to climate change.* Bogota. ISBN 978-958-98840-1-0: UN DC- COLOMBIA.

UNFCCC, CM (1994). *https://unfccc.int.* Retrieved 03/10/2019, from https://unfccc.int/files/essential_background/background_publications_htmlpdf/application/pdf/convsp.pdf

Costa, P. C. (2007). Adaptation to climate change in Colombia. *Revsita de Ingenierias, UNIANDES* (26), 74-80.

Daniel Licandro, O., Alvarado-Peña, LJ, Sansores Guerrero, EA, & Navarrete Marneou, JE (2019). Corporate Social Responsibility: Towards the formation of a typology of definitions. *Venezuelan Management Magazine* , 3-14; ISSN: 1315-9984; Redalyc: https://www.redalyc.org/articulo.oa?

Delgado, GC, De Luca Zuria, A., Vázquez Zentella, V., ., :, & . (2015). *Urban adaptation and mitigation of climate change in Mexico Title.* Mexico: Center for Interdisciplinary Research in Sciences and Humanities, National Autonomous University of Mexico.

EIU, EI (2010). *Latin American Green Cities Index, A comparative evaluation of the ecological impact of the main cities of Latin America.* Munich, Germany: Siemens, Green City Index.

SHE, E. and. (2013). City-level-climate-change-adaptation-strategies-the-case-of-quito-ecuador. *ELLA, Environmental Management* , 1-6.

FAO. (2021). *FAO in Colombia* . Food: going from losses to solutions: http://www.fao.org/colombia/noticias/detail-events/en/c/1238132/

Fonfria, S. (04/17/2017). *Local Energy and Climate Change Agency Murcia.* http://www.um.es/documents/3456781/5197411/Presentacion+Plan+Adaptacion+Murcia_ALEM.pdf/9611c9b6-d7f1-4c3c-8f5f-9d35789c246e

FONFRIA, S. (04/17/2017). *Local Energy and Climate Change Agency Murcia.* http://www.um.es/documents/3456781/5197411/Presentacion+Plan+Adaptacion+Murcia_ALEM.pdf/9611c9b6-d7f1-4c3c-8f5f-9d35789c246e

Forman, RT, Sperling, D., Bissonette, JA, Clevenger, AP, Cutshall, CD, Dale, VH, . . . Winter, T. C. (2003). *Road Ecology: Science and Solutions.* Island Press.

Galindo, LM, Samaniego, J., Alatorre, JE, Carbonell, JF, Reyes, O., & Sanchez, L. (2015). *Eight theses on climate change and sustainable development in Latin America.* Santiago de Chile: UN.

Garabiza Castro, B. d., Sánchez Guerrero, JF, & Casanova Montero, AR (2017). The internalization of business externalities in Ecuador. *Res non verba (Guayaquil)* , 47-64, University of Guayaquil.

García, LA (2018). Externalities and road culture. Phenomena around car use in Xalapa, Veracruz, Mexico. *Cleavages, social sciences magazine. ISSN: 2395-9495* , 171-187; https://doi.org/10.25009/clivajes-rcs.v0i9.2539.

GEA21, G. d. (2017). *State of the art of Climate Action Plans, Climate Action Plan 2050 of Donostia / San Sebastián.* San Sebatian, Spain: GEA21. http://www.aclima.eus/wp-content/uploads/2017/09/Estado-del-arte-de-los-Planes-de-Acci%C3%B3n-del-Clima-V3.pdf

Government of Chile, M. d. (2017). *NATIONAL CLIMATE CHANGE ACTION PLAN 2017-2022.* Santiago de Chile: Government of Chile.

GOVERNMENT OF CHILE, MINSITERY OF THE ENVIRONMENT. (2017). *NATIONAL CLIMATE CHANGE ACTION PLAN 2017-2022.* Santiago de Chile: Government of Chile.

Government of the Republic of Colombia. (2019). *National circular economy strategy. Closing of materials cycles, technological innovation, collaboration and new business models.* http://www.andi.com.co/Uploads/Estrategia%20Nacional%20de%20EconA%CC%83%C2%B3mia%20Circular-2019%20Final.pdf_637176135049017259.pdf

Heras, B. P. (2015). ADAPTATION TO CLIMATE CHANGE IN THE EUROPEAN UNION: LIMITS AND POTENTIALITIES OF A MULTILEVEL POLICY. *ELECTRONIC JOURNAL OF INTERNATIONAL STUDIES* (24), 1-29.

Hernandez, SR, Fernando, FC, & Pilar, BL (2014). *INVESTIGATION METHODOLOGY.* MEXICO: MAC GRAW HILL.

HERNANDEZ, SR, Fernando, FC, & Pilar, BL (2014). *INVESTIGATION METHODOLOGY*. MEXICO: MAC GRAW HILL.

HERRERO, AC, NATENZON, C., & LORENA, MM (17 of 11, 2018). *Social vulnerability, threats and risks to climate change in the Greater Buenos Aires Agglomerate.* Buenos Aires: CIPPEC Publications. https://www.lanacion.com.ar/2084753-buenos-aires-lider-contra-el-cambio-climatico

Herrero, AC, Natenzón, C., & Lorena, MM (17 of 11, 2018). *Social vulnerability, threats and risks to climate change in the Greater Buenos Aires Agglomerate.* Buenos Aires: CIPPEC Publications. https://www.lanacion.com.ar/2084753-buenos-aires-lider-contra-el-cambio-climatico

IBARRA, ML (2016). *Social vulnerability in Tijuana due to hydrometeorological events. Case study: Colony October 3.* Tijuana, Mexico: CFN, Colegio de la Frontera Norte.

IDEAM, I. d. (2015). *New Climate Change Scenarios for Colombia 2011-2100, Scientific Tools for Decision Making.* Bogotá: IDEAM publications.

IDEAM, I. d. (2016). *Third National Communication on Climate Change.* Bogota: IDEAM, Institute of Hydrology, Meteorology and Environmental Studies.

INI, L. (2016). *renewable-energies.com.* https://www.energias-renovables.com/panorama/vaxjo-la-ciudad-mas-verde-de-europa-20161007

IPCC, P.I. (2014). www.ipcc.ch. In *Climate Change 2014, Impacts, Adaptation and Vulnerability* (pp. 35-94). United Kingdom and New York: IPCC.

Laurance, W.F., Clements, G.R., Sloan, S., O'Connell, C.S., Mueller, N.D., Goosem, M., . . . Burgues Arrea, I. (2014). A global strategy for road construction. *Nature* , 229-232.

Laurance, W.F., Goosem, M., & Laurance, S.G. (2009). Impacts of roads and linear clearings on tropical forests. *Trends in Ecology & Evolution* , v. 24, no. 12, .

López-Guzmán Guzmán, TJ (2009). Socioeconomic development of rural areas based on community tourism. A case study in Nicaragua. . *Rural development notebooks* , 81-97.

Magrin, GJ-P. (2014). *Climate Change 2014: Impacts, Adaptation, and Vulnerability. Part B:Regional Aspects. Contribution of Working Group II to the Fifth Assessment Report for IPCC.* new york: UN.

Margulis, S. (2016). *Vulnerability and adaptation of Latin American cities to climate change.* Santiago: UN, ECLAC, Economic Commission for Latin America and the Caribbean.

MARGULIS, S., & CEPAL, CE (2016). *Vulnerability and adaptation of Latin American cities to climate change.* Santiago: UN.

María Margarita Bedoya-V., AA-A.-V. (2018). Wildlife accidents in the urban road network of five cities in the Aburrá Valley (Antioquia, Colombia). *CONSERVATION* , 335-348.

Medellin, A. d. (2015). *Comprehensive strategy for the management of Climate Change.* Medellin.

Messmer, T. A. (2008). Deer–vehicle collision statistics and mitigation information: online sources. *Human-Wildlife Conflicts* , 131-135.

Minenvironment. (2020). *List of specific environmental impacts within the scope of environmental licensing.*

MinAmbiente, & CAEM, CA (2016). *Comprehensive Atlantic Territorial Climate Change Management Plan 2040.* Barranquilla.

MINAMBIENTE, M. d. (2017). *Comprehensive Territorial Climate Change Management Plan of the Department of Santander 2030.* Bogota DC: MINAMBIENTE.

Molina, M., Jose, S., & Julia, C. (2017). *Climate change, causes, effects and solutions.* Mexico: Economic Culture Fund.

Monroy, MC-L. (2015). Rate of roadkill of wildlife on the San Onofre–María la Baja road, Colombian Caribbean. *Magazine of the Colombian Association of Biological Sciences* , 88 - 95.

Montenegro Montero, HM (2018). *Wildlife Run Over on Via Mamatoco - Minca, Santa Marta, Colombian Caribbean.* Santa Marta: Magdalena University.

Moroney, A. (2018). *The use of spatial and temporal analysis in the maintenance of road mortality mitigation measures for wildlife in Ireland.* . Stockholm: KTH ROYAL INSTITUTE OF TECHNOLOGY SCHOOL OF ARCHITECTURE AND THE BUIL ENVIRONMENT.

United Nations. (2021). *United Nations Colombia* . Moving from food losses and waste to solutions: https://nacionesunidas.org.co/noticias/actualidad-colombia/pasando-de-perdidas-y-desperdicios-de-alimentos-pda-a-soluciones/

United Nations. (May 18, 2021). *Responsible production and consumption: why they are important.* https://www.un.org/sustainabledevelopment/es/wp-content/uploads/sites/3/2016/10/12_Spanish_Why_it_Matters.pdf

Niemeyer, O., & Costa, L. (2018). *Brasilia, the smart city of the past.* Brasilia: Smart.City_Lab.

Novillo, r. N., Olmedo, MP, Perez, Y., & Rojas, PY (2018). *Approaches to the Study of the relationship between cities and climate change.* Quito, Ecuador: FLACSO.

ORTIZ, BL (2017). *The Externalities.* www.economia.unam.mx: http://www.economia.unam.mx/profesores/blopez/valoracion-externalidades.pdf

Pabón, CJ (2018). Climate change in Colombia. *Geography Reviews, National University of Colombia .*

Global Pact. (May 18, 2021). *Companies and organizations before SDG 12 .* https://www.pactomundial.org/2019/11/sector-privada-ante-ods-12/

Pilar, R. (2013). *FAO.* http://www.fao.org/3/a-i3388s.pdf

Republic of Colombia . (2019). *Gazette of the Congress, Senate and Chamber.* Total amendment to the text proposed for first debate to bill number 301 of 2018: http://leyes.senado.gov.co/proyectos/images/documentos/Textos%20Radicados/Ponencias/2019/gaceta_357.pdf

Rojas, AD (2016). *Road development in Colombia and the impact of fourth generation roads.* Bogota: UMNG, Nueva Granada Military University.

Rozas, P. &. (2004). Infrastructure development and economic growth: conceptual review. *ECLAC .*

Sánchez, RR (2013). *Urban Responses to Climate Change in Latin America.* Santiago de Chile: ECLAC, ECLAC-IAI, Economic Commission for Latin America and the Caribbean.

Seijas, AE-Q. (2013). Vertebrate mortality on the Guanare-Guanarito highway, Portuguesa state, Venezuela. *Journal of Tropical Biology* , 1619-1636.

Sengupta, S. (03/25/2019). Copenhagen, Can a city cancel its greenhouse gas emissions? *The clarin .*

Smathers Jr, W. (2001). The socioeconomic impacts of wildlife-vehicle collisions. . *Wildlife and highways.* , twenty-one.

Sustainable, D. (1986). Sustainable Development Goals. . *Food and Agriculture Organization: Rome, Italy.*

Stasiukynas, DC-W. (2021). Roads to the sea: study on the impact of wild vertebrates and surrounding ecosystems in two road corridors in Colombia. *Science Technology Society Trilogy* , 13 (24).

Superintendence of home public services. (2019). *Solid waste final disposal report 2018.* Bogotá: Superintendence of household public services.

Tsegaye, B., Jaiswal, S., & Jaiswal, A. (2021). Food waste biorefinery: Pathway toward circular bioeconomy. *Foods, 10* (6), 1174. https://doi.org/https://doi.org/10.3390/foods10061174

UNICC, UC (12, 2017). *Case Study: Paradigms of Mobility.* https://ciudadarquitecturamedioambiente.files.wordpress.com/2015/09/movilidad-curitiba-alex-levet.pdf

UNICC, CRISTOBAL COLON UNIVERSITY. (12 of 2017). *Case Study: Paradigms of Mobility.* https://ciudadarquitecturamedioambiente.files.wordpress.com/2015/09/movilidad-curitiba-alex-levet.pdf

Unit, E.I. (2010). *Index of Green Cities of Latin America.* siemens. https://www.siemens.com/press/pool/de/events/corporate/2010-11-lam/study-latin-american-green-city-index_spain.pdf

Valencia, AM (2015). *.ayto-valencia.es.* http://www.ayto-valencia.es/ayuntamiento/Energias.nsf/0/7ABD14C45FA34ECAC1257ED7002C2A96/$FILE/Plan%20de%20accion%20ambiental.pdf?OpenElement&lang=1

VALENCIA, AM (2015). *.ayto-valencia.es.* http://www.ayto-valencia.es/ayuntamiento/Energias.nsf/0/7ABD14C45FA34ECAC1257ED7002C2A96/$FILE/Plan%20de%20accion%20ambiental.pdf?OpenElement&lang=1

World Bank Group. (2018). *What a waste 2.0. A global snapshot of solid waste management to 2050.* https://openknowledge.worldbank.org/handle/10986/30317

CAPITULO 4.

MORTALIDAD DE FAUNA SILVESTRE POR EFECTO VEHICULAR EN LA TRONCAL CENTRAL (RUTA NACIONAL 45 A08) EN EL TRAMO BUCARAMANGA (SECTOR PALENQUE-LA CEMENTO) A SAN ALBERTO (CESAR)

CHAPTER 4.

WILDLIFE MORTALITY DUE TO VEHICULAR EFFECT ON THE CENTRAL TRUNK (NATIONAL ROUTE 45 A08) IN THE BUCARAMANGA SECTION (PALENQUE-LA CEMENTO SECTOR) TO SAN ALBERTO (CESAR)

Carolina Hernandez Contreras

Biologist. M.Sc. in Environmental Sciences and Technologies

Research professor of the GIECSA Research Group, assigned to the Environmental Engineering program of the Technological Units of Santander

chernandez@correo.uts.edu.co

Yenifer Katherine Flórez Contreras

Environmental Engineer

Student Participant of the GIECSA research group, assigned to the Environmental Engineering program of the Technological Units of Santander

ykflorez1212@gmail.com

Jessica Viviana Vega Sánchez

Environmental Engineer

Student Participant of the GIECSA research group , assigned to the Environmental Engineering program of the Technological Units of Santander

jviviana.s@hotmail.com

ODS:

SUMMARY

Currently, our region presents socioeconomic growth and at the same time development of road infrastructure that leads to environmental problems and within this the running over of wildlife, presenting great changes in the ecosystems that surround the roads. Taking into account the above, the objective of the following book chapter is to describe the critical points that the Central trunk (national route 45 A 08) presents between the Bucaramanga section (Palenque -

La Cemento sector) and San Alberto (Cesar) regarding the wildlife mortality due to vehicle collision and propose strategies that lead us to meet the goals of SDG 15.

The study was carried out in order to determine the mortality of individuals due to vehicle collision and to know a preliminary percentage that shows how many animals die due to this situation, as well as to determine the impact that the loss of that species generates on the ecosystem. According to their order of importance by deaths, wild species were found, firstly the faunal group of birds, secondly, reptiles and finally mammals. Likewise, the research proposed alternative solutions to reduce and mitigate the collision of the species that predominate in this area.

It is expected that with this research, it will be possible to prioritize the conservation of biodiversity present on the route by entities such as INVIAS, ANI, environmental authorities and entities in charge of the roads. Furthermore, in the sections identified as critical, it is recommended to strengthen mitigation measures. since they are intended to make drivers reduce their speed or local fauna to avoid the most dangerous sections.

INTRODUCTION

Currently there is great social and economic growth which requires a modification of the road network that offers multiple benefits to the economy, communication and recreation of the community. Due to the modernization of this infrastructure, a great impact on the faunal biodiversity present where these projects are carried out was evident.

There are a series of change factors that directly and indirectly affect wild populations through the development of road systems (Forman y otros, 2003). The road network within or near natural areas has been identified as one of the main causes of habitat loss and fragmentation (Laurance y otros, Impacts of roads and linear clearings on tropical forests., 2009). Currently, it is estimated that 25 million kilometers of roads will have been built by 2050, 90% of which will be in developing countries (Laurance y otros, 2014).

A central problem is evident, such as the running over of wildlife, which was addressed in the research project carried out on the central trunk road (national route 45 a08) in the section Bucaramanga (Palenque-La Cemento Sector) to San Alberto (Cesar) where the critical points of the road were identified and how the ecosystems in said area are affected by the loss of species that are impacted.

According to the research carried out on the Toluviejo highway in the Ciénaga la Caimanera by Ossa & Galván, 2015 was used as a reference to carry out the current investigation, where some factors that affect the accident rate of species such as vehicular flow, width of the road, the behavior of the species, the vegetation cover, the absence of signage. (Montenegro Montero, 2018)

Based on this information, the methodology used to carry out the research was proposed. The project was carried out in three phases, the first being the collection of primary and secondary information, to obtain the primary information it was carried out through monitoring in the study area, the secondary information was obtained through the review of bibliography of public entities and private.

Subsequently, the analysis of the data collected during the tours was carried out to determine the critical points of the study route, in turn the most threatened species in this fragmented ecosystem were identified. Likewise, the population dynamics between the species were established and how the habitat they inhabit is disturbed by its absence.

The data collected in this research project demonstrate the negative effect of the Central trunk (national route 45 A 08) between the Bucaramanga section (Palenque - La Cemento sector) - San Alberto (Cesar) on the fauna, a problem that requires the articulation of the different entities such as INVIAS, ANI, environmental authorities, educational institutions, as well as the communities residing on the road.

IDENTIFIED PROBLEM

The development of road infrastructure is a fundamental element in the socioeconomic growth of a country, which requires that modernization be provided in the physical infrastructure of Colombia, presenting a benefit to the community because it facilitates their mobilization, reduces the cost and time of transportation of the products, facilitating the entry and exit of these goods to places that are difficult to access. However, as economic development progresses and the construction of dual carriageways advances, they cause adverse impacts on the natural dynamics of the wild species that They inhabit ecosystems immersed in the network. (Rozas, 2004) (López-Guzmán Guzmán, 2009).

The running over of wildlife on roads is one of the global problems that comes from the expansion and construction of roads because they interfere with natural dynamics. Roads and vehicular traffic impact wildlife in four ways: 1) deterioration in the quantity and quality of the ecosystem they inhabit 2) difficult access to food for wildlife 3) increased mortality due to species being run over 4) fragmentation of wildlife communities, on the other hand, the species that are highly affected are the largest ones since they have a smaller number of individuals, are more mobile and have few reproduction rates. (Smathers Jr, 2001), (Moroney, 2018), (Messmer, 2008)

Santander has various roads that help with good connectivity between departments. The central trunk (national route 45A 08) contains 501 km in length which crosses various ecosystems, allowing for a diversity of wildlife in the Bucaramanga-San Alberto (Cesar) section.), called first-order road, the increase in vehicular flow is evident, generating inappropriate use and abuse in speed, quantity and frequency; Added to this is the lack of environmental awareness on the part of some drivers, who prefer to cause injuries or even death to animals found on the road.

It is important to highlight that the competent environmental authorities do not require construction companies to identify and evaluate the negative effects that road projects produce on wildlife due to being run over, as well as the effect that is caused on the surrounding ecosystems. Likewise, the management plans do not contemplate strategies to prevent and mitigate the impact on the fauna present, such as fencing systems, signage, reflectors, underpasses, and overpasses. Demonstrating our central problem, it is necessary to formulate the following question to provide you with the best solution strategy. What is the mortality behavior of wildlife due to being run over on the road called Troncal Central (national route 45 A 08) between the Bucaramanga section (Palenque - La Cemento sector) - San Alberto (Cesar) and what is the contribution to the SDG? fifteen?

The protection and conservation of natural and environmental wealth is the responsibility of everyone and all activities. This responsibility can be assumed through direct action or through shared participation, that is, after joining common efforts to help improve or transform the situation. Commercial agents can help with the strengthening of environmental aspects, being protagonists of actions, promoting change in others and supporting those who take action for the benefit of all.

Companies focus on responding with maximum quality to their commercial niche, but today this is not enough for the vision of quality of life that, from a brand or product, society recognizes as an agent of well-being. Today it is expected that, in addition to commercial or productive quality, goods and resources transmit social responsibility, in which their contribution to their clients, not due to the specific consumption marketed, but after their commercial positioning generates added values, which, without additional cost, they provide well-being to the buyer or consumer. Within this social responsibility, a fundamental column, environmental social responsibility, from which interest and contribution is evident in a product, either through its own action, or through coordination with others, for the protection, conservation and use of natural and environmental resources. , in two fundamental ways: in the sustainable use of resources to generate its marketable product and in its contribution to the environmental environment in investments and positions that are not linked to the commercialization of its product.

The analysis of deaths and impacts of fauna due to transportation systems on road axes (García, 2018) shows a space in which commercial activities impact nature, indirectly and without the intention of its deterioration, this is an understandable position. but that must be evident within commercial and productive activities, so that all those responsible (guilty or malicious) take the initiative to remedy the damage to nature that their commercial interest entails. All commerce that moves by land transportation on this axis has a responsibility to promote actions that protect the flora and fauna that the road axis generates.

The flora can be compensated, after replanting and replacement, to recover green coverage in the future, if not equal to that affected by the roads, if at least in progressive strengthening that makes them environmentally usable. The fauna requires in situ and immediate actions, in which mechanisms are sought that adapt reality to the ecosystem needs of nature and a strategy is established to mitigate the damage that, the road, land transportation and the goods that are there. They market, generate goods and ecosystem services (Rojas, 2016).

METHODOLOGY

The first activity that was developed was the diagnosis of the species found in the ecosystems fragmented by the central trunk (National Route 45 A08) in the Bucaramanga section (Palenque - La Cemento sector) to San Alberto (Cesar). For this, the study area was delimited using the QGis tool, consecutively the type of ecosystem that predominates on the road was defined with the help of studies from competent authorities such as IDEAM, IGAC, IAVH, and the Holdridge methodology.

12 samples were carried out in a period of two months (May-June 2019) in which each one had 4 routes during the day (morning/afternoon), obtaining a total of 48 routes; where it was necessary to travel two observers on a motorcycle at a speed of 20 km/h. The study period is

carried out according to the authors' availability time. The registration of the species was carried out in situ, taking into account geographical location and date of observation, in the same way a photographic record was kept to carry out the analysis and identification of the bodies of the animals.

For the characterization of the animals registered in the section, it was carried out according to the definitions of the taxonomic categories of each individual that are established in the red books defined by the Alexander Von Humboldt Institute (IAVH). As support, the iNaturalist application was used, which made it possible to immediately know specific characteristics of the species. Along with photographic evidence, the coordinates of the place where the animal was found run over were attached. A survey was applied to the community located in the area of influence, which strengthened the identification of the most predominant species and those most threatened by roadkill.

Then, the population dynamics were analyzed taking into account the mortality rate of wildlife due to being run over on the central trunk (National Route 45 A08) in the Bucaramanga section (Palenque - La Cemento sector) to San Alberto (Cesar). The roadkill rate by taxonomic class was calculated taking into account the registered roadkill individuals, the days surveyed (48) and the length of the route (107 km) (Seijas, 2013), (Monroy, 2015) as seen in the following formula:

$$\text{Hit and run rate (TA)} = \text{N hits} / (\text{N km} * \text{N traveled}).$$

To analyze the impact of being run over by each type of individual, each section was evaluated, identifying which one has the greatest number of cases. Depending on the geographical location where the road is located, shapefiles from the SIAC (Colombian Environmental Information System) were used, where the ecosystems and biomes present in the project area were obtained. Taking into account the types of biomes, the rate of roadkill was analyzed for each class of registered individuals, in order to establish how many species are lost for each biome.

Finally, we analyzed how we could contribute to SDG 15 with this study and formulated various environmental strategies that mitigate the impact of wildlife on the central trunk (National Route 45A) in the Bucaramanga-San Alberto (Cesar) section. After diagnosing and analyzing mortality due to wildlife being run over, strategies were proposed that help reduce, mitigate and protect the species present in the ecosystems in order to minimize the impacts generated on this trunk.

RESULTS AND DISCUSSION

For the diagnosis of the species, the identification of the study area was first carried out, corresponding to 107 km of national highway, included between the department of Santander (starting in the La Cemento sector E:1104783;N:1283475) and ending at the limit with Cesar (San Alberto with E:1076115; N1349504) the altitude range goes from 100 to 111 meters above sea level, with temperature >24°C, the route crosses the municipalities of Bucaramanga, Rionegro, Playón, La Esperanza and ends at the limit of the municipality of Saint Albert. This route is a single carriageway with 2 lanes where there is no signage for the conservation and protection of fauna. According to Holdridge's (1978) classification, the area comprises three life zones: Tropical very dry forest (bms-T), tropical humid forest (bh-T), and premontane rainforest (bp/pm).

Registry of roadkill species.

At the end of the monitoring in the central trunk (national route 45 to 08) in the Bucaramanga section (Palenque-La Cement sector) to San Alberto (Cesar), 147 individuals were recorded being run over in 48 routes, which were carried out in a period of 2 months. (May-June) of the year 2019 "Table 1", of which 119 were identified with their genus or species, on the contrary, 28 species were not identified due to the conditions in which the corpses were found, but they were classified according to some characteristics typical of a faunal group (mammals, birds, reptiles, insects).

Table 1. Species hit by roadkill in 107 km of the central trunk road (national route 45 a08)

Family	Common name	Scientific name	No of Runovers
	Mammalia class		
Didelphidae	common chucha common opossum	*Didelphis marsupiali*	twenty-one
	domestic dog	*Canis lupus familiaris*	4
Myrmecophagidae	Honeyeater or Anteater	*Mexican tamandua*	2
Sciurus	Squirrel	*Sciurus granatensis*	1
Chiropterans	Bat	Chiroptera	1
Unidentified			3
Bird Class			

Cathartidae	Gallinazo	*Coragyps atratus*	fifteen
Emberizidae	Canary	*Sicalis flaveola*	7
Thraupidae	silver spike toche	*Ramphocelus dimidiatus*	2
Thraupidae	Palmero Tile	*Thraupis palmarum*	2
Picidae	Woodpecker	*Colaptes punctigula*	1
Columbidae	reddish dove	*Columbina talpaco*	2
Anatidae	Duck	*Anatinae anas*	1
T **rochilidae**	hummingbird	*Chlorostilbon mellisugus*	1
Psittacidae	green parakeet	*Psittacara holochlorus*	3
Psittacidae	spectacled parakeet	*Forpus conspicillatus*	1
Tyrannidae	common mahi mahi	*Pseudocolopteryx flaviventris*	1
Thraupidae	Tile bird	*Thraupis episcopus*	1
Thraupidae	Black and white Benteveo	*Tyrannus tyrannus*	1
Turdidae	May	*Turdus ignobilis*	3
Fringillidae	White-winged Goldfinch	*Spinus psaltria*	1
Unidentified			6
		Reptilia Class	
Colubridae	chili tail	*Micrurus mipartitus*	2
	common iguana	*Delicate iguana*	
Iguanidae			
	green Iguana	*Iguana iguana*	9
Colubridae	Tiger, toche, flying	*Spilotes pullatus*	3
Colubridae	Loggerhead snake	*Mastigodryas boddaerti*	1
Colubridae	Green jacket	*Leptophis ahaetulla*	3
Viperidae	Size x, dick tail	*Bothrops asper*	1
Elapidae	Coral	*Micrurus dumerilii*	1
Dipsadidae	false coral	*Erythrolamprus mimus*	10
Dipsadidae	Common Snail	*Sibon nebulatus*	2

Dipsadinae	Male coral (false)	*Pseudoboa neuwiedii*	2
Salamandridae	common lizard	*salamander salamander*	1
Unidentified			9
	Amphibious Class		
Ranidae	Frog	*Hypsiboas pugnax*	1
Jesters	Toad	*rhinella marina*	2
Gastropoda Class			
Achatinidae	African snail	*Achatina fulica*	7
	Class Insecta		
Without Identification (Actinote)			
		Class Arachnida	
Araneae	Spider	YEAH	1
		Class Myriapoda	
Annelida	Worm	*Hylesia nigricans*	1
Total			**146**

Source: Authors

With the above information it can be deduced that birds were the group with the highest incidence, with 48 individuals equivalent to 33%, showing that the species with the highest records was *Coragyps atratus* . The second group with the largest number of individuals belongs to the class of reptiles with 44 individuals representing 30%, the most affected species was the *Erythrolamprus mimus* with 23%, and the Iguanidae family represented 19% of the affected individuals. Regarding the class of mammals, it was categorized as the third group with the highest incidence of roadkill with 33 individuals, which corresponds to 22%, where Didelphis *marsupialis* represents 64% of the total number of mammals recorded, finally insects represented 6. 8%, gastropods with 4.8%, amphibians with 2% and finally the classes with the lowest incidence of roadkill were myriapods and arachnids with 0.7% each.

Deforestation and desertification are anthropogenic activities that pose serious challenges to sustainable development and have affected the lives and livelihoods of millions of people. Natural ecosystems are vitally important to sustaining life on Earth and play a key role in the fight against climate change. Each species has an ecological niche in the habitat where it lives, for this reason its environment is highly affected if there is an absence of it since it is a fundamental component of the biodiversity and ecological balance of the ecosystem , animals play a determining role, They are protagonists of a large part of the phenomena and processes that guarantee adequate conditions for life.

Climatic factors influence the food supply and reproductive season, which is related to the greater or lesser number of road kills; Likewise, the greater or lesser coverage of foraging areas is related to the time of year. At the end of the rains and the beginning of the drought, the production of wild fruits is greater. Phenological seasonality also notably affects the composition, structure and dynamics of the ecosystem. (Bauni, 2017).

Based on the information collected and the statistical data, the species with the highest accident rate are emphasized, highlighting the importance that this plays in the environment and in its population.

An individual that is highly threatened in our study area corresponds to the vulture (*Coragyps atratus*), a scavenger bird, as observed in studies carried out on the urban road network in the Aburra Valley (María Margarita Bedoya-V., 2018). These birds approach the road to eat the animals that are decomposing. Furthermore, another factor that affects the mortality problem of this species is the organic waste thrown by people who circulate on the roads. In the same way, the greatest cause of bird collision is caused when they take small grains of sand from the edge of the road to have better digestion of the seeds. It was also observed that these birds build their nests in the tops of trees, looking affected by the passage of heavy-duty vehicles, which rub the branches, destabilizing the nest boxes until they fall to the asphalt and subsequently cause the death of the individuals. Since there is an absence of these species in the ecosystem, the ecological balance will be disturbed because seed dispersal will not be carried out on a regular basis.

Opossums (*Didelphis marsupiali*) have a wide habitat range and record high levels of migration between populations compared to other small mammals. Its high mobility and nocturnal habits and small size may be related to the poor visibility to detect them on the road and avoid being run over, this explains why it is the second species with the highest records. It is also found that this species is reported in studies carried out in road accidents in Cali-Buenaventura roads (Stasiukynas, 2021), as is also reported for the urban network of the Aburrá Valley (María Margarita Bedoya-V., 2018), on the coastal roads of the Gulf of Mexico, loss of the same species is also reported but in smaller quantities (Canales-Delgadillo, 2020). It is known that this species is important in the ecosystem since it helps the regeneration of the forest because they are seed dispersers. They thus generate positive impacts for the environment and for human populations .

Another species of the Mammalia class that is vulnerable to being run over is the anteater, which feeds mainly on insects (ants, termites and bees). It has generally nocturnal and arboreal habits, for which it has a prehensile tail that facilitates its movement between trees. Its locomotion on land is rather slow, so crossing a road easily puts it in danger of being run over. This phenomenon, along with hunting and habitat degradation, constitute the main threats that affect their populations (IAVH, 2017).

For the second phase of the research, the study of population dynamics was carried out and the collision rate (TA) was calculated according to the formula of (Monroy, 2015) (Seijas, 2013)0.028 km/day of vehicular collisions, which according to the authors is low impact.

Next, each taxonomic class is analyzed with roadkill records taking into account its geographical location (Longitude, Latitude, Height) in order to determine the section that has the highest rate of roadkill and thus analyze the impact of roadkill for each type of vehicle. individual, this was evaluated by sections, identifying which one has the greatest number of cases.

Birds are the class with the highest mortality rate on the road, for this it was determined that the section with the highest rate of collisions was section 2 (Rionegro-Playón) with 0.0046ind/day/km. and 24 records, followed by section 3 (Playón-Cáchira) with 11 records and a collision rate of 0.00214ind/day/km. "Illustration 1".

Figure 1. Registration number per section.

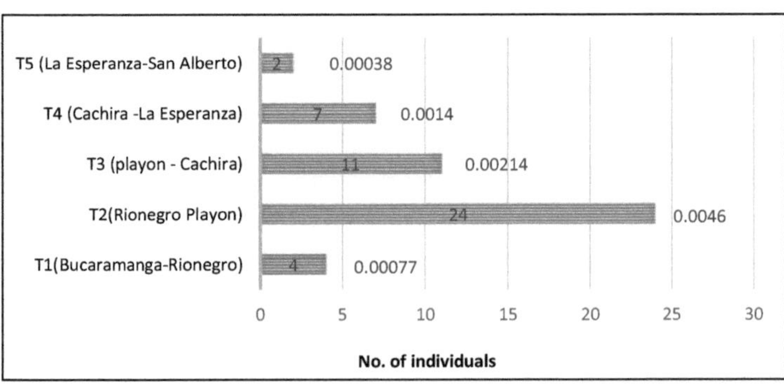

Source: Authors

The reptiles have a record of being run over by 44 individuals, in which it was determined that section 2 (Rionegro-Playón) has the highest rate of being run over with 0.0041 with 21 records, followed by section 3 (Playón-Cáchira) with a rate of 0.00233ind/day/km. and 12 records, in the third instance is section 4 (Cáchira-La Esperanza) with 11 records that corresponds to a rate of 0.00214ind/day/km. Finally, section 1 (Bucaramanga, Rionegro) and section 5 (La Esperanza-San Alberto) do not have records of roadkill species. "Illustration 2".

Illustration 2. Registration number per section.

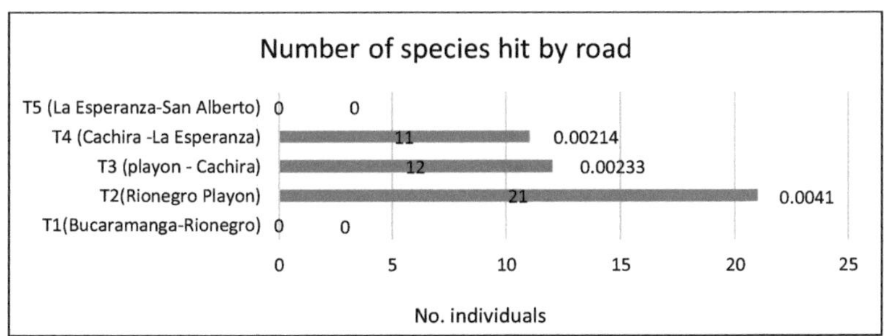

Source: Authors

On the other hand, 33 mammals were recorded on the road, in which it was established that section 2 (Rionegro-Playón) has the highest rate of run over with 0.00311ind/day/km and 16 records, followed by section 4 (Cáchira -La Esperanza) with 12 records and a collision rate of 0.00233ind/day/km, later section 3 (Playón-Cáchira). It is important to mention that section 1 (Bucaramanga-Rionegro) does not have records of roadkill species "Illustration 3".

Illustration 3. Registration number per section.

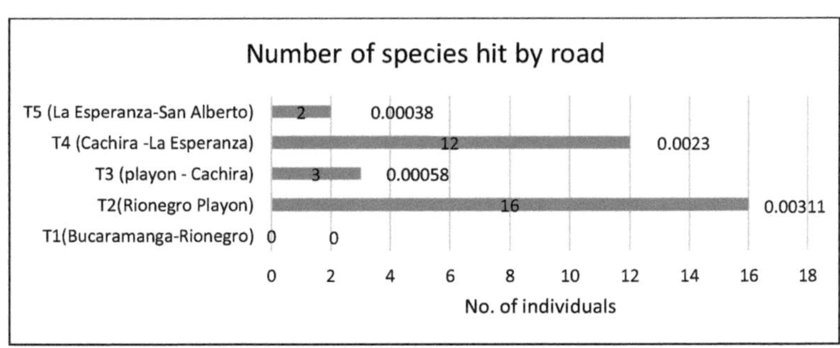

Source: Authors

The Insecta class, for its part, has a record number of 10 individuals run over, it was established that section 3 (Playón-Cáchira) has a higher rate of run over with 0.00311ind/day/km and 8 records, followed by the section 2 (Rionegro-Playón) with 2 records and a collision rate of 0.00038ind/day/km, meaning that section 1 (Bucaramanga-Rionegro), section 4 (Cáchira-La Esperanza) and section 5 (La Esperanza -San Alberto) do not have records of roadkill species "Illustration 4".

Illustration 4. Registration number per section.

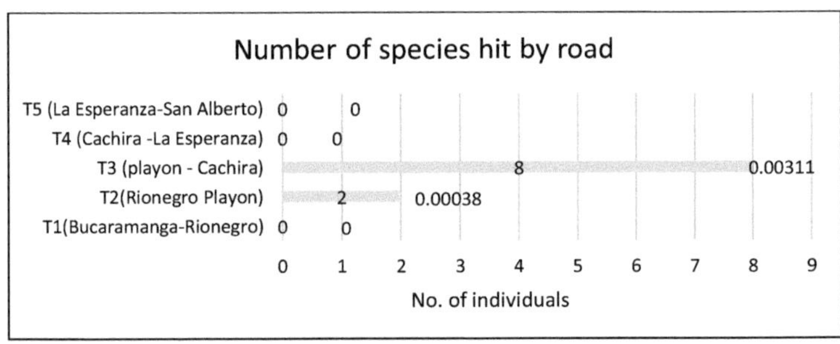

Source: Authors

In the class of gastropods, a total of 7 individuals were recorded during the investigation, in which two sections with a roadkill rate were found, which was section 2 (Rionegro-Playón) with a rate of 0.0012ind/day/km and 6 records, meanwhile section 3 (Playón-Cáchira) with 1 record and a collision rate of 0.00019 ind/day/km "Illustration 5".

Illustration 5. Registration number per section.

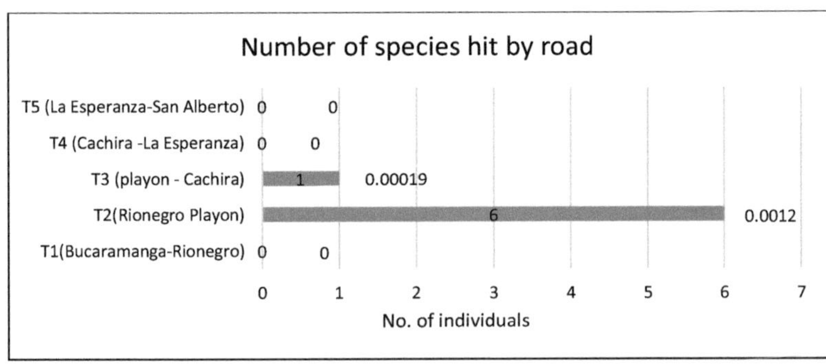

Source: Authors

In the class of amphibians, a total of 3 individuals were recorded during the investigation, in which two sections with a roadkill rate were found, which was section 2 (Rionegro-Playón) with a rate of 0.00039ind/day/km and 2 records, meanwhile section 3 (Playón-Cáchira) with 1 record and a collision rate of 0.00019 ind/day/km "Illustration 6".

Illustration 6. Registration number per section.

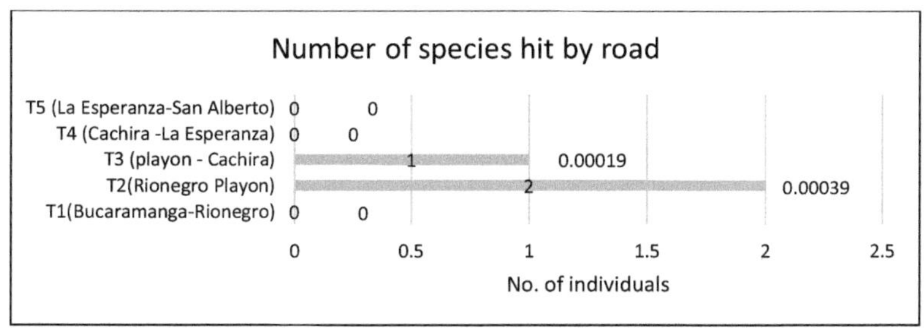

Source: Authors

Myriapods have a single record of an individual being run over in section 2 (Rionegro-Playón) with a roadkill rate of 0.00019ind/day/km "Illustration 7".

Illustration 7. Registration number per section.

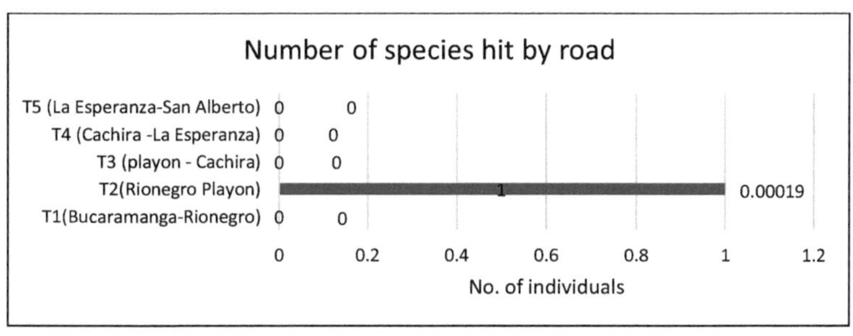

Source: Authors

Likewise, a single individual is recorded for the class of arachnids and it is found roadkill on section 2 (Rionegro-Playón) with a roadkill rate of 0.00019ind/day/km "Illustration 8".

Illustration 8. Registration number per section.

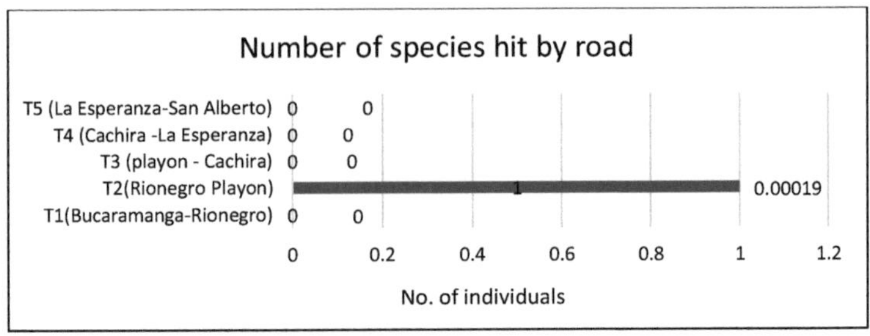

Source: Authors

The total AT of the present study is 0.02647ind/km/day, it is relatively low when compared with studies carried out on the San Onofre – María la Baja road, Colombian Caribbean where the total AT was 0.328ind/km/day, in the same way It is relatively low when compared to the studies carried out in Portuguesa, Venezuela, which is 0.1393ind/km/day (Seijas, 2013).

Each species fulfills an ecological niche in the habitat where it lives, for this reason its environment is highly affected if there is an absence of it since it is a fundamental component of the biodiversity and ecological balance of the ecosystem, animals play a determining role , are leading actors in a large part of the phenomena and processes that guarantee adequate conditions for life.

Climatic factors influence the food supply and reproductive season, which is related to the greater or lesser number of road kills; Likewise, the greater or lesser coverage of foraging areas is related to the time of year. At the end of the rains and the beginning of the drought, the production of wild fruits is greater. Phenological seasonality also notably affects the composition, structure and dynamics of the ecosystem. (Bauni, Anfuso, & Schivo, 2017).

CONTRIBUTION TO THE SDGS

To prevent, halt and reverse the degradation of ecosystems around the world, the United Nations has declared the Decade for Ecosystem Restoration (2021-2030). This globally coordinated response to habitat loss and degradation will focus on developing the political will and capacity to restore humans' relationship with nature. Likewise, it is a direct response to the warning of science, as expressed in the Special Report on Climate Change and Land of the

Intergovernmental Panel on Climate Change, to the decisions adopted by all Member States of the United Nations. United in the Rio conventions on climate change and biodiversity and the United Nations Convention to Combat Desertification. (Sostenible, 1986)

By 2030, ensure the conservation of mountain ecosystems, including their biological diversity, to improve their capacity to provide essential benefits for sustainable development and Take urgent and significant measures to reduce the degradation of natural habitats, halt the loss of biological diversity and, by 2020, protect threatened species and prevent their extinction (Sostenible, 1986).

To comply with the sustainable development objectives, especially taking into account the previously described goals of SDG 15 in these critical points of wildlife mortality due to vehicular effect on the Central Trunk (National Route 45 a08) in the Bucaramanga Section (Palenque Sector -La Cemento) to San Alberto (Cesar), it is proposed to implement structural or non-structural strategies that provide the individual with better security for their movement.

Within the structural strategies, underground or aerial wildlife passages are recommended in order to allow a connection between the fauna and the habitat that have been or will be fragmented by the construction of the roads.

The aerial passages are built between trees intended especially for mammals such as monkeys, squirrels, opossums, sloth bears, among others. It is based on the development of elevated platforms that allow the continuous movement of animals between the trees.

As for the underground fauna passages, they are of the box type where they are intended for all types of vertebrates (small, medium, large) without ruling out the use of these structures by other types of animals such as reptiles, amphibians, among others, for this it is necessary to delimit with a mesh that contains plant material in order to guide access to the tunnel.

As for the signaling measures, they are those that report the presence of fauna species in the area, since they have not been implemented on this road. The purpose of these signs is to warn drivers of the possible collision of animals that usually cross the road (Bonds, 2001).

Finally, we could also take into account non-structural strategies since they contemplate the reduction of speed taking into account that the majority of wildlife collisions occur due to the high speed they travel, possibly increasing the timely response capacity of drivers. of vehicles that travel these roads.(Castillo-R, 2015)

Understanding the results of this document, from environmental economics, represents an important contribution to business understanding of the environmental problems that its activity of exchanging goods and services generates on nature and the environment. Recognizing that every marketable product, in some way, generates externalities on nature, is a key step to achieve the sustainability of production and the execution of actions that mitigate, recover and restore the ecosystem aspects that are affected.

An externality (ORTIZ, 2017), defined with a simple concept, is the damage that the person responsible for an activity generates on a third party, with whom they do not have direct contact, but which deteriorates their living environment conditions. The product transported in a vehicle that, without malicious intent, runs over an animal is responsible for the deterioration it causes to that individual, to its living environment, to its ecosystem. This action constitutes an externality not assumed by the person responsible for the marketed product and not assumed in costs and indications to avoid or recover good environmental conditions.

When companies assume their environmental social responsibility (Daniel Licandro y otros, 2019), they must assume administrative and strategic actions that allow them to identify their interventions, potential and real, on nature so that in both possibilities, a position is assumed to avoid these interventions, mitigate their effects by recovering the damages and losses that it generates. Land commercial transportation has a high level of environmental impact, any company or commercial activity that requires land mobility of its products is immersed in externalities and I call for assuming responsibility to internalize them (include in the cost structure, the economic amount of recover the environmental damage generated indirectly).

Understanding that mitigation, recovery and restoration actions of nature require a specialist and not every productive system can be an expert in its environmental scope, it is important that strategies be promoted that integrate administrative managers and biological and environmental experts to execute processes at favor of nature and the common living environment.

Internalizing damage to nature, such as road accidents, can be achieved in two simple ways: one, including in the commercial image of the product, aspects that demonstrate its identity with the environment and its initiative to respond to its actions with the environment. ; and two, promote actions, our own and those of others, for the protection of the affected fauna, after in-situ adaptation to isolate the affecting agent from environmental resources, goods and services. (Garabiza Castro, Sánchez Guerrero, & Casanova Montero, 2017).

CONCLUSIONS AND RECOMMENDATIONS

Marsupials (Possums) are the group with the largest recorded individuals. The frequency of roadkill can be attributed to the abundance of this species in the region and its behavior of feeding on other roadkill. By direct observation it was established that due to their nocturnal habit they are dazzled by the light of vehicles, leaving them immobile and therefore more susceptible to being run over.

The factor of bird mortality due to vehicle collisions on the Central trunk (national route 45 A 08) between the Bucaramanga section (Palenque - La Cemento sector) -San Alberto (Cesar) on fauna, can refer to the diversity of vegetation. arboreal and shrubby on the sides of the trunk, minimizing the availability of passage places for these species. Likewise, by direct observation it was determined that birds generally cross or remain on the road in search of food. Likewise, it was established that birds die when colliding with cars and not directly being run over.

For the study area, there is a total TA value of 0.02647ind/km/day. This can be considered an indirect method of population assessment. Animal deaths due to road accidents are analogous to "captures per unit." of effort" where the "capture" is the number of animals run over and the "effort" is a direct function of vehicular traffic. If there is an increase in vehicular traffic, a decrease or stagnation in the TA can be interpreted as a decrease in the population of the analyzed species.

During the sampling, guavas (Psidium guajava) and mangoes (Mangifera indica) could be found on the road, which were food for birds and mammals, turning out to be more susceptible to being run over or hit, due to their low flight and long stay. on the road.

As for section 2 (Rionegro-playón), it was determined as the most critical point, throughout the entire trajectory of the investigation. Since it has tree and bush vegetation on both sides of the road, therefore, many animals may feel attracted to stay in these places making use of the resources, because there are no barriers that prevent the species from going out onto the road.

.

In conclusion, the data collected in this research demonstrate the negative effect of the Central trunk (national route 45 A 08) between the Bucaramanga section (Palenque - La Cemento sector) -San Alberto (Cesar) on the fauna, a problem that requires articulation of the different entities such as INVIAS, ANI, environmental authorities, educational institutions, as well as the communities residing on the road.

This research exposes the need to urgently implement measures to reduce the collision of species in the central trunk 45 to 08. It is expected that the proposed measures will be implemented and once implemented these strategies should be evaluated in the long term. its effectiveness.

It is expected that with this research, it will be possible to prioritize the conservation of biodiversity present on the route by entities such as INVIAS, ANI, environmental authorities and entities in charge of the road. Furthermore, in the sections identified as critical, it is recommended to strengthen mitigation measures. since they are intended to make drivers reduce their speed or local fauna to avoid the most dangerous sections.

Likewise, it is recommended that the management plans required by ANLA or the competent entities have strategies that mitigate the impacts generated by the construction or modification of any road network.

It is also recommended for future studies to take into account the months of the year or times of the year with respect to the vehicular flow, identifying greater and lesser vehicular flow to have a reference for loss of biodiversity due to road traffic.

For future studies, if possible, take into account cameras on the road since not all hit-and-run individuals can be detected; some may fall off the road after the impact, be injured and seek refuge outside it, therefore, the Actual mortality may be much higher.

REFERENCES

EEA, A.E. (2015). *Climate Change and Cities.* Madrid, Spain: European Environment Agency.

European Environment Agency, EEA. (2018). *10 Case Studies How Europe is adapting to climate change.* https://climate-adapt.eea.europa.eu/about/climate-adapt-10-case-studies-online.pdf

MAYOR OF IBAGUE. (2016). *alcaldiadeibague.gov.co.* http://www.alcaldiadeibague.gov.co/portal/admin/archivos/publicaciones/2016/14024-PLA-20160502.pdf

Ibague Mayor's Office. (2016). *alcaldiadeibague.gov.co.* http://www.alcaldiadeibague.gov.co/portal/admin/archivos/publicaciones/2016/14024-PLA-20160502.pdf

Amaya, CC, Tavera, CN, Avila, AM, Alvarado, PJ, and Vesga, Á. P. (2019). Methodological Proposal for Evaluation of the Level of Vulnerability to Climate Change in Urban Environments. Commune 13, Bucaramanga, Santander. *17th LACCEI International Multi-Conference for Engineering, Education, and Technology* (pp. 1-10). Boca Raton - USA: LACCEI.

AMAYA, CC, and HERNANDEZ, CC (04/15/2018). *Pascualbravo.edu.co/cintexpb* . Pascualbravo.edu.co: http://www.pascualbravo.edu.co:5056/cintexpb/index.php/cintex/article/view/301

AMB, A.M. (2011). *Mobility Master Plan for Bucaramanga 2011-2030.* Bucaramanga: AMB.

Arboleda, A. (2008). Manual for the evaluation of environmental impacts of projects, works or activities., (pp. 75 - 86). Medellin.

Barcelano, A. (2020). *Barcelona for Climate, Ecology, Urban Planning, Infrastructure and Mobility.* https://www.barcelona.cat: https://www.barcelona.cat/barcelona-pel-clima/es/barcelona-responde/acciones-concretas

Bárcena, IA, Samaniego, J., Peres, W., and Alatorre, JE (2020). *The emergency of climate change in Latin America and the Caribbean: do we continue waiting for the catastrophe or do we take action?* Washington DC: United Nations - ECLAC. LC/PUB.2019/23-P. ISBN 9789211220315.

BARTON, JR (2009). Adaptation to climate change in city-region planning. *Magazine of Geography Norte Grande, 43* (1), 5-30.

Bauni, V. A. (2017). Wildlife mortality due to road kill in the Atlantic forest of Alto Paraná, Argentina. *Ecosystems Magazine* , 54-56.

IDB, BI (2016). *Evaluation of the IDB Emerging and Sustainable Cities Initiative.* Washington, DC: IDB.

IDB, Inter-American Development Bank. . (2106). *Methodological guide of the Emerging and Sustainable Cities Program.* Washington DC USA: IDB.

Bogota. (2015). *District plan for adaptation and mitigation to climate variability and change.* Bogota.

Bonds, B. 2. (2001). Wildlife habitat mitigation. *In: Wildlife and highways:seeking solutions to an ecological and socio-economic dilemma. 7th Annual Meeting of The Wildlife Society. Nashville, Tennessee* , 70-72.

Breil, M., Downing, C., Kazmierczak, A., Mäkinen, K., & Romanovska, L. (2018). *Social vulnerability to climate change in European cities – state of play in policy and practice.* Bologna, Italy. https://doi.org/10.25424/CMCC/SOCVUL_EUROPCITIES: European Topic Center on Climate Change impacts, Vulnerability and Adaptation (ETC/CCA).

Bucaramanga, AM (2012). *Territoiral Planning Plan 2012-2027.* Bucaramanga: Mayor of Bucaramanga, http://www.concejodebucaramanga.gov.co/pot-2012-2027/tomo01.pdf.

Buenos Aires, G. d. (2017). *Provincial Report, Adaptation of the Sustainable Development Goals in the city of Buenos Aires.* Buenos Aires, Argentina: Government of the city of Buenos Aires.

Cali, AM (2020). *Municipal plan for adaptation and mitigation to climate change.* Cali. Cauca Valley.

Canales-Delgadillo, JP-C.-J.-P.-P.-M. (2020). Traffic deaths on the Gulf of Mexico coastal highway: how many and which species of wildlife are being lost? *Mexican Biodiversity Magazine* , 91.

Castillo-R, JC-M.-G. (2015). Wildlife mortality due to vehicle collision in a sector of the Pan-American Highway between Popayán and Patía. *Scientific Bulletin. Museum Center. Museum of Natural History,* , 207 -219.

Chavarro, PM, Garcia, GA, Garcia, PJ, Pabón, JD, Prieto, RA, & Ulloa, CA (2013). *Preparing for the future, threats, risks, vulnerability and adaptation to climate change.* Bogota. ISBN 978-958-98840-1-0: UN DC- COLOMBIA.

UNFCCC, CM (1994). *https://unfccc.int.* Retrieved 03/10/2019, from https://unfccc.int/files/essential_background/background_publications_htmlpdf/application/pdf/convsp.pdf

Costa, P. C. (2007). Adaptation to climate change in Colombia. *Revsita de Ingenierias, UNIANDES* (26), 74-80.

Daniel Licandro, O., Alvarado-Peña, LJ, Sansores Guerrero, EA, and Navarrete Marneou, JE (2019). Corporate Social Responsibility: Towards the formation of a typology of definitions. *Venezuelan Management Magazine* , 3-14; ISSN: 1315-9984; Redalyc: https://www.redalyc.org/articulo.oa?

Delgado, GC, De Luca Zuria, A., Vázquez Zentella, V., ., :, and . (2015). *Urban adaptation and mitigation of climate change in Mexico Title.* Mexico: Center for Interdisciplinary Research in Sciences and Humanities, National Autonomous University of Mexico.

EIU, EI (2010). *Latin American Green Cities Index, A comparative evaluation of the ecological impact of the main cities of Latin America.* Munich, Germany: Siemens, Green City Index.

SHE, E. and. (2013). City-level-climate-change-adaptation-strategies-the-case-of-quito-ecuador. *ELLA, Environmental Management* , 1-6.

FAO. (2021). *FAO in Colombia* . Food: going from losses to solutions: http://www.fao.org/colombia/noticias/detail-events/en/c/1238132/

Fonfria, S. (04/17/2017). *Local Energy and Climate Change Agency Murcia.* http://www.um.es/documents/3456781/5197411/Presentacion+Plan+Adaptacion+Murcia_ALEM.pdf/9611c9b6-d7f1-4c3c-8f5f-9d35789c246e

FONFRIA, S. (04/17/2017). *Local Energy and Climate Change Agency Murcia.* http://www.um.es/documents/3456781/5197411/Presentacion+Plan+Adaptacion+Murcia_ALEM.pdf/9611c9b6-d7f1-4c3c-8f5f-9d35789c246e

Forman, RT, Sperling, D., Bissonette, JA, Clevenger, AP, Cutshall, CD, Dale, VH, . . . Winter, T. C. (2003). *Road Ecology: Science and Solutions.* Island Press.

Galindo, LM, Samaniego, J., Alatorre, JE, Carbonell, JF, Reyes, O., and Sanchez, L. (2015). *Eight theses on climate change and sustainable development in Latin America.* Santiago de Chile: UN.

Garabiza Castro, B. d., Sánchez Guerrero, JF, and Casanova Montero, AR (2017). The internalization of business externalities in Ecuador. *Res non verba (Guayaquil)* , 47-64, University of Guayaquil.

García, LA (2018). Externalities and road culture. Phenomena around car use in Xalapa, Veracruz, Mexico. *Cleavages, social sciences magazine. ISSN: 2395-9495* , 171-187; https://doi.org/10.25009/clivajes-rcs.v0i9.2539.

GEA21, G. d. (2017). *State of the art of Climate Action Plans, Climate Action Plan 2050 of Donostia / San Sebastián.* San Sebatian, Spain: GEA21. http://www.aclima.eus/wp-content/uploads/2017/09/Estado-del-arte-de-los-Planes-de-Acci%C3%B3n-del-Clima-V3.pdf

Government of Chile, M. d. (2017). *NATIONAL CLIMATE CHANGE ACTION PLAN 2017-2022.* Santiago de Chile: Government of Chile.

GOVERNMENT OF CHILE, MINSITERY OF THE ENVIRONMENT. (2017). *NATIONAL CLIMATE CHANGE ACTION PLAN 2017-2022.* Santiago de Chile: Government of Chile.

Government of the Republic of Colombia. (2019). *National circular economy strategy. Closing of materials cycles, technological innovation, collaboration and new business models.* http://www.andi.com.co/Uploads/Estrategia%20Nacional%20de%20EconA%CC%83%C2%B3mia%20Circular-2019%20Final.pdf_637176135049017259.pdf

Heras, B. P. (2015). ADAPTATION TO CLIMATE CHANGE IN THE EUROPEAN UNION: LIMITS AND POTENTIALITIES OF A MULTILEVEL POLICY. *ELECTRONIC JOURNAL OF INTERNATIONAL STUDIES* (24), 1-29.

Hernandez, SR, Fernando, FC, and Pilar, BL (2014). *INVESTIGATION METHODOLOGY.* MEXICO: MAC GRAW HILL.

HERNANDEZ, SR, Fernando, FC, and Pilar, BL (2014). *INVESTIGATION METHODOLOGY .* MEXICO: MAC GRAW HILL.

HERRERO, AC, NATENZON, C., and LORENA, MM (17 of 11, 2018). *Social vulnerability, threats and risks to climate change in the Greater Buenos Aires Agglomerate.* Buenos Aires: CIPPEC Publications. https://www.lanacion.com.ar/2084753-buenos-aires-lider-contra-el-cambio-climatico

Herrero, AC, Natenzón, C., and Lorena, MM (17 of 11, 2018). *Social vulnerability, threats and risks to climate change in the Greater Buenos Aires Agglomerate.* Buenos Aires: CIPPEC Publications. https://www.lanacion.com.ar/2084753-buenos-aires-lider-contra-el-cambio-climatico

IBARRA, ML (2016). *Social vulnerability in Tijuana due to hydrometeorological events. Case study: Colony October 3.* Tijuana, Mexico: CFN, Colegio de la Frontera Norte.

IDEAM, I. d. (2015). *New Climate Change Scenarios for Colombia 2011-2100, Scientific Tools for Decision Making.* Bogotá: IDEAM publications.

IDEAM, I. d. (2016). *Third National Communication on Climate Change.* Bogota: IDEAM, Institute of Hydrology, Meteorology and Environmental Studies.

INI, L. (2016). *renewable-energies.com.* https://www.energias-renovables.com/panorama/vaxjo-la-ciudad-mas-verde-de-europa-20161007

IPCC, P.I. (2014). www.ipcc.ch. In *Climate Change 2014, Impacts, Adaptation and Vulnerability* (pp. 35-94). United Kingdom and New York: IPCC.

Laurance, W.F., Clements, G.R., Sloan, S., O'Connell, C.S., Mueller, N.D., Goosem, M., . . . Burgues Arrea, I. (2014). A global strategy for road construction. *Nature* , 229-232.

Laurance, W.F., Goosem, M., and Laurance, S.G. (2009). Impacts of roads and linear clearings on tropical forests. *Trends in Ecology & Evolution* , v. 24, no. 12, .

López-Guzmán Guzmán, TJ (2009). Socioeconomic development of rural areas based on community tourism. A case study in Nicaragua. . *Rural development notebooks* , 81-97.

Magrin, GJ-P. (2014). *Climate Change 2014: Impacts, Adaptation, and Vulnerability. Part B:Regional Aspects. Contribution of Working Group II to the Fifth Assessment Report for IPCC.* new york: UN.

Margulis, S. (2016). *Vulnerability and adaptation of Latin American cities to climate change.* Santiago: UN, ECLAC, Economic Commission for Latin America and the Caribbean.

MARGULIS, S., and CEPAL, CE (2016). *Vulnerability and adaptation of Latin American cities to climate change.* Santiago: UN.

María Margarita Bedoya-V., AA-A.-V. (2018). Wildlife accidents in the urban road network of five cities in the Aburrá Valley (Antioquia, Colombia). *CONSERVATION* , 335-348.

Medellin, A. d. (2015). *Comprehensive strategy for the management of Climate Change.* Medellin.

Messmer, T. A. (2008). Deer–vehicle collision statistics and mitigation information: online sources. *Human-Wildlife Conflicts* , 131-135.

Minenvironment. (2020). *List of specific environmental impacts within the scope of environmental licensing.*

MINAMBIENTE, M. d. (2017). *Comprehensive Territorial Climate Change Management Plan of the Department of Santander 2030.* Bogota DC: MINAMBIENTE.

MinAmbiente, and CAEM, CA (2016). *Comprehensive Atlantic Territorial Climate Change Management Plan 2040.* Barranquilla.

Molina, M., Jose, S., and Julia, C. (2017). *Climate change, causes, effects and solutions.* Mexico: Economic Culture Fund.

Monroy, MC-L. (2015). Rate of roadkill of wildlife on the San Onofre–María la Baja road, Colombian Caribbean. *Magazine of the Colombian Association of Biological Sciences* , 88 - 95.

Montenegro Montero, HM (2018). *Wildlife Run Over on Via Mamatoco - Minca, Santa Marta, Colombian Caribbean.* Santa Marta: Magdalena University.

Moroney, A. (2018). *The use of spatial and temporal analysis in the maintenance of road mortality mitigation measures for wildlife in Ireland.* . Stockholm: KTH ROYAL INSTITUTE OF TECHNOLOGY SCHOOL OF ARCHITECTURE AND THE BUIL ENVIRONMENT.

United Nations. (2021). *United Nations Colombia* . Moving from food losses and waste to solutions: https://nacionesunidas.org.co/noticias/actualidad-colombia/pasando-de-perdidas-y-desperdicios-de-alimentos-pda-a-soluciones/

United Nations. (May 18, 2021). *Responsible production and consumption: why they are important.* https://www.un.org/sustainabledevelopment/es/wp-content/uploads/sites/3/2016/10/12_Spanish_Why_it_Matters.pdf

Niemeyer, O., and Costa, L. (2018). *Brasilia, the smart city of the past.* Brasilia: Smart.City_Lab.

Novillo, r. N., Olmedo, MP, Perez, Y., and Rojas, PY (2018). *Approaches to the Study of the relationship between cities and climate change.* Quito, Ecuador: FLACSO.

ORTIZ, BL (2017). *The Externalities.* www.economia.unam.mx: http://www.economia.unam.mx/profesores/blopez/valoracion-externalidades.pdf

Pabón, CJ (2018). Climate change in Colombia. *Geography Reviews, National University of Colombia* .

Global Pact. (May 18, 2021). *Companies and organizations before SDG 12* . https://www.pactomundial.org/2019/11/sector-privada-ante-ods-12/

Pilar, R. (2013). *FAO.* http://www.fao.org/3/a-i3388s.pdf

Republic of Colombia . (2019). *Gazette of the Congress, Senate and Chamber.* Total amendment to the text proposed for first debate to bill number 301 of 2018: http://leyes.senado.gov.co/proyectos/images/documentos/Textos%20Radicados/Ponencias/2019/gaceta_357.pdf

Rojas, AD (2016). *Road development in Colombia and the impact of fourth generation roads.* Bogota: UMNG, Nueva Granada Military University.

Rozas, P. &. (2004). Infrastructure development and economic growth: conceptual review. *ECLAC* .

Sánchez, RR (2013). *Urban Responses to Climate Change in Latin America.* Santiago de Chile: ECLAC, ECLAC-IAI, Economic Commission for Latin America and the Caribbean.

Seijas, AE-Q. (2013). Vertebrate mortality on the Guanare-Guanarito highway, Portuguesa state, Venezuela. *Journal of Tropical Biology* , 1619-1636.

Sengupta, S. (03/25/2019). Copenhagen, Can a city cancel its greenhouse gas emissions? *The clarin* .

Smathers Jr, W. (2001). The socioeconomic impacts of wildlife-vehicle collisions. . *Wildlife and highways.* , twenty-one.

Sustainable, D. (1986). Sustainable Development Goals. . *Food and Agriculture Organization: Rome, Italy.*

Stasiukynas, DC-W. (2021). Roads to the sea: study on the impact of wild vertebrates and surrounding ecosystems in two road corridors in Colombia. *Science Technology Society Trilogy* , 13 (24).

Superintendence of home public services. (2019). *Solid waste final disposal report 2018.* Bogotá: Superintendence of household public services.

Tsegaye, B., Jaiswal, S., and Jaiswal, A. (2021). Food waste biorefinery: Pathway toward circular bioeconomy. *Foods, 10* (6), 1174. https://doi.org/https://doi.org/10.3390/foods10061174

UNICC, UC (12, 2017). *Case Study: Paradigms of Mobility.* https://ciudadarquitecturamedioambiente.files.wordpress.com/2015/09/movilidad-curitiba-alex-levet.pdf

UNICC, CRISTOBAL COLON UNIVERSITY. (12 of 2017). *Case Study: Paradigms of Mobility.* https://ciudadarquitecturamedioambiente.files.wordpress.com/2015/09/movilidad-curitiba-alex-levet.pdf

Unit, E.I. (2010). *Index of Green Cities of Latin America.* siemens. https://www.siemens.com/press/pool/de/events/corporate/2010-11-lam/study-latin-american-green-city-index_spain.pdf

Valencia, AM (2015). *.ayto-valencia.es.* http://www.ayto-valencia.es/ayuntamiento/Energias.nsf/0/7ABD14C45FA34ECAC1257ED7002C2A96/$FILE/Plan%20de%20accion%20ambiental.pdf?OpenElement&lang=1

VALENCIA, AM (2015). *.ayto-valencia.es.* http://www.ayto-valencia.es/ayuntamiento/Energias.nsf/0/7ABD14C45FA34ECAC1257ED7002C2A96/$FILE/Plan%20de%20accion%20ambiental.pdf?OpenElement&lang=1

World Bank Group. (2018). *What a waste 2.0. A global snapshot of solid waste management to 2050.* https://openknowledge.worldbank.org/handle/10986/30317

GENERAL BIBLIOGRAPHY

Arbuze, Ivosn & Huacon, Gianella (2018): "The evolution of productivity and quality in goods and services companies", Observatory of Latin American Economy Magazine, (January 2019). Online: https://www.eumed.net/rev/oel/2019/01/empresas-bienes-servicios.html

Combeller, C. R. (1993). The New Scenario: The Culture of Quality and Productivity in Companies. Jalisco, Mexico.: ITESO.

Gomez, O.E. (2018). Strategic cost management, a competitiveness tool. Revsita Espacios, , 4-14, Vol. 39 (No. 32).

Isaac, GC, Gomez, BJ, & Díaz, AS (2017). THE INTEGRATION OF ENVIRONMENTAL MANAGEMENT TOOLS AS A SUSTAINABLE PRACTICE IN ORGANIZATIONS. University and Society Magazine, 27-36; Vol 17 No 4.

UN, UN (2015). Resolution to adopt the 2030 Agenda for Sustainable Development. Paris: UN.

uts
Unidades Tecnológicas de Santander
¡Lo hacemos posible!
UDI
UNIVERSIDAD DE INVESTIGACIÓN Y DESARROLLO
CO2

Printed by Books on Demand GmbH, Norderstedt / Germany